Quality Management and Operations Research

Quality Management and Operations Research
Understanding and Implementing the Nonparametric Bayesian Approach

Nezameddin Faghih, Ebrahim Bonyadi, and Lida Sarreshtehdari

CRC Press
Taylor & Francis Group
Boca Raton London New York

CRC Press is an imprint of the
Taylor & Francis Group, an **informa** business

First edition published 2021
by CRC Press
6000 Broken Sound Parkway NW, Suite 300, Boca Raton, FL 33487-2742

and by CRC Press
2 Park Square, Milton Park, Abingdon, Oxon, OX14 4RN

© 2021 Nezameddin Faghih, Ebrahim Bonyadi, and Lida Sarreshtehdari

CRC Press is an imprint of Taylor & Francis Group, LLC

The right of Nezameddin Faghih, Ebrahim Bonyadi, and Lida Sarreshtehdari to be identified as authors of this work has been asserted by them in accordance with sections 77 and 78 of the Copyright, Designs and Patents Act 1988.

Reasonable efforts have been made to publish reliable data and information, but the author and publisher cannot assume responsibility for the validity of all materials or the consequences of their use. The authors and publishers have attempted to trace the copyright holders of all material reproduced in this publication and apologize to copyright holders if permission to publish in this form has not been obtained. If any copyright material has not been acknowledged please write and let us know so we may rectify in any future reprint.

Except as permitted under U.S. Copyright Law, no part of this book may be reprinted, reproduced, transmitted, or utilized in any form by any electronic, mechanical, or other means, now known or hereafter invented, including photocopying, microfilming, and recording, or in any information storage or retrieval system, without written permission from the publishers.

For permission to photocopy or use material electronically from this work, access www.copyright.com or contact the Copyright Clearance Center, Inc. (CCC), 222 Rosewood Drive, Danvers, MA 01923, 978-750-8400. For works that are not available on CCC please contact mpkbookspermissions@tandf.co.uk

Trademark notice: Product or corporate names may be trademarks or registered trademarks and are used only for identification and explanation without intent to infringe.

Library of Congress Cataloging-in-Publication Data
Names: Faghih, Nezameddin, author. | Bonyadi, Ebrahim, author. |
Sarreshtehdari, Lida, author.
Title: Quality management and operations research : understanding and
implementing the nonparametric Bayesian approach / Nezameddin Faghih,
Ebrahim Bonyadi, and Lida Sarreshtehdari.
Description: First edition. | Boca Raton, FL : CRC Press, 2021. |
Includes bibliographical references and index.
Identifiers: LCCN 2020049335 (print) | LCCN 2020049336 (ebook) |
ISBN 9780367744908 (hardback) | ISBN 9781003158141 (ebook)
Subjects: LCSH: Quality control—Statistical methods. |
Bayesian statistical decision theory.
Classification: LCC TS156 .F33 2021 (print) | LCC TS156 (ebook) |
DDC 658.5/62—dc23
LC record available at https://lccn.loc.gov/2020049335
LC ebook record available at https://lccn.loc.gov/2020049336

ISBN: 978-0-367-74490-8 (hbk)
ISBN: 978-1-003-15814-1 (ebk)

Typeset in Times
by codeMantra

This book, published during the COVID-19 outbreak, is dedicated to all students worldwide, especially those who could not afford even the least expensive computing devices to attend their online classes.

Contents

Foreword ix
Preface xi
Acknowledgments xv
Authors xvii

Chapter 1 Introduction 1

1.1 Need for Quality 1
1.2 Quality Management 2
    1.2.1 Quality Management Parameters 3
1.3 Quality Determinants 4
1.4 Factors Affecting Reliability 6

Chapter 2 Quality and Reliability 9

2.1 Some Remarkable Properties of Survival Data 10
2.2 Important Functions for Assessing Failure Time 10
    2.2.1 Cumulative Distribution Function 11
    2.2.2 Probability Density Function 11
    2.2.3 Survival Function 11
    2.2.4 Hazard Function 12
    2.2.5 Quantile Function 14
2.3 Censor 15
2.4 Accelerated Lifetime Tests 16
    2.4.1 Accelerated Lifetime Tests Models 17
    2.4.2 Lifetime-Stress Relationship 18
2.5 Bayesian Approach 18
2.6 Markov Chain Monte Carlo Method 20
    2.6.1 Monte Carlo Approach 20
        2.6.1.1 Monte Carlo Integration 21
        2.6.1.2 Importance Sampling 22
    2.6.2 Markov Chain 23
        2.6.2.1 Definitions 23
        2.6.2.2 Chain Structure 25
        2.6.2.3 Limiting Distribution of Chain 25
    2.6.3 Metropolis–Hastings Algorithm 28
        2.6.3.1 Gibbs Sampling Method 29
        2.6.3.2 Some Features of the Gibbs Sampling Method 30
2.7 Slice Sampling 31

Chapter 3 Dirichlet Process 33

3.1 Dirichlet Distribution.. 33
       3.1.1 Remarkable Properties of the Dirichlet Distribution........................ 38
3.2 Dirichlet Process.. 43
3.3 Pólya's Urn Model... 49
       3.3.1 Pólya's Urn Process... 49
       3.3.2 Blackwell–MacQueen Urn Scheme.. 50
3.4 Dirichlet Process and Clustering Issue... 52
       3.4.1 Chinese Restaurant Process... 54

Chapter 4 Nonparametric Bayesian Approach in Accelerated Lifetime Tests..... 59

4.1 Dirichlet Process Mixture Models 60
       4.1.1 Mixture Models... 60
4.2 Log-linear Regression in the Nonparametric Problem......................... 63
4.3 Determining the Base Distribution and the Precision Parameter 65
4.4 Hierarchical Model of the Dirichlet Process 66
4.5 Bayesian Computation 67
4.6 Model Fitting 69
       4.6.1 First Stage: Updating Θ_i... 70
       4.6.2 Second Stage: Updating Θ_j^*.. 72
       4.6.3 Third Stage: Updating ξ... 73
       4.6.4 Fourth Stage: Updating β... 73
       4.6.5 Fifth Stage: Updating the Distribution of Failure-Time 73

Chapter 5 Illustrative Examples and Results....................................... 77

5.1 Empirical Distribution Function.. 77
5.2 Dirichlet Process Weibull Mixture Model 78
       5.2.1 Determining Base Distribution... 78
5.3 Assessing the Model and Simulation.. 80
       5.3.1 Updating (α_i, λ_i).. 80
       5.3.2 Updating $(\alpha_j^*, \lambda_j^*)$.. 83
       5.3.3 Updating ϕ, γ, and μ... 85
              5.3.3.1 Updating ϕ... 85
              5.3.3.2 Updating γ... 86
              5.3.3.3 Updating μ... 87
              5.3.3.4 Updating β... 88
5.4 Illustrative Examples 89

Appendix A: Guide to Proofs... 97

Appendix B: R Programming Codes 101

References 115

Index... 119

Foreword

As Covid-19 continues to decimate many aspects of our lives, rebuilding and redesigning is critically important. Before the Covid-19 epidemic, we tended to think of Bayesian modeling as limited to engineering and mathematical applications. However, as others will discover, the Bayesian approach provides a tool that resonates across multiple disciplines. This includes such diverse applications as economics, management, entrepreneurship, design of six sigma adoption, and marketing roll-out.

With all the disruptions and change that Covid-19 has brought to our societies, successfully going forward is now more highly dependent on selecting the best option. Selecting 'the best option from within a limited grouping' may not seem new. In fact, since the end of the Great Depression in 1939, we have become accustomed to looking for the best opportunity, among multiple choices. In the immediate aftermath of the economic impact from the Covid-19 destruction, we will be considerably less well-off.

Our response will require careful planning and adapting to better modeling of the choices that are actually viable. A Bayesian approach is an exceptionally useful tool in the new world of difficult and seemingly less-related choices that emerge from the pandemic. It behoves each of us to familiarize ourselves with the techniques the Bayesian approach offers.

The following pages in this book could not have come at a better time! With a little patience, this book can become a valuable tool for the broader population when making difficult decisions.

Victoria Hill
Meknes, Morocco

Preface

The growing demand of consumers to increase the quality and guarantee of products has always been one of the main concerns of manufacturers in the world. The importance of this issue is so crucial that it can be said it is the only factor that creates competition among business owners.

Undoubtedly, a quality product always wins the competition. Therefore, trying to produce a reliable and quality product is equivalent to trying to attract the sales market.

Achieving this depends on managing quality, design, beauty, efficiency, packaging, and transportation. But what is important is to produce a high-quality product that lasts for a long time.

Technological advancements and improving the quality of people's lives through the use of industrial achievements have created intense competition among manufacturers, products, and services. In the production of strategic products, if a company fails to deliver the quality products/services in compliance with the desired standards, it will soon exit the competition.

The rapid advances in technology, the development of modern products, the intensification of global competition, and the increasing expectations of customers have put new pressure on high-quality product manufacturers. Customers expect to buy products that are reliable and secure. Systems, machines, and devices must be capable of operating at high probability under standard conditions for a specified period.

Hence, applying robust methods to analyze lifetime data entails in-depth knowledge and wide skills from various metric instruments and mathematical tools. The most relevant way in the statistics in which many parametric and nonparametric methods have been developed is the reliability analysis (also known as survival analysis).

In the first chapter, we first describe some features of reliability data, and since we need some functions to interpret these features and to obtain some of the features of these data, we briefly introduce some of these functions. In the next sections of Chapter 2, we will refer to censored data issue and will study accelerated lifetime tests, as one of the most important discussions in reliability data analysis. Since the data studied in this book are accelerated by an accelerator, we also describe the relationship between lifetime data and stress, and finally, both the Bayesian approach and the Markov chain Monte Carlo (MCMC), which will be utilized throughout this study, will be discussed in this chapter. Also, the slice sampling method, which is one of the sampling methods in this book, will be briefly reviewed.

On the other hand, in statistical inference, there are data from a known distribution with an unknown parameter $\theta \in \Theta \subset \mathcal{R}^d$, which in most cases d is finite and the main goal is to estimate the parameter/parameters with less error.

One of the methods used in point estimation of parameters is the Bayesian method. In this method, the parameter is considered as a variable and it is also necessary to choose a prior distribution for the parameter. Using Bayes' law, the posterior distribution of the parameter is computed on the condition of the data, and finally, the researcher estimates the parameter using the posterior distribution. This is generally called the parametric Bayesian method.

But in practice, there are many cases where the distribution of data is not known, and unlike the former state, where the goal was the parameter estimation, in such cases, the objective is predicting the distribution of data. It should be noted that the Bayesian approach can also be used in this method. In this case, the Bayesian method is called nonparametric Bayes. The major difference between parametric and nonparametric Bayes is their parameter dimension (or equivalently the dimension of distribution function). Therefore, in the parametric Bayes, the dimension of the parameter is finite, and in the nonparametric Bayes, an infinite number of parameters will come out.

Most studies assume that the parameters have known prior distributions, such as normal distributions. However, this assumption is sometimes a restrictive idea. Because unknown parameters, considered as a random variable, may come from another distribution, such as a skewed, multimodal, or a heavy-tail distribution. Therefore, the use of a nonparametric background is an effective approach to removing this limitation.

Since in the nonparametric Bayesian approach the dimension of the parameter is infinite, it is assumed the parameter to be a random stochastic process rather than a random variable. A stochastic process covers all family of distributions and can be considered as a suitable prior distribution for the parameters. Due to the complexity of the computations in the nonparametric Bayesian method, choosing the stochastic process as a prior for all unknown parameters is an important issue.

The selected stochastic process should be flexible and the calculations within must come easier and faster. In other words, the considered stochastic process must be conjugate. Among these processes are the Dirichlet process and Polya's tree. Ferguson, therefore, introduced the Dirichlet process as the prior of all distribution functions in 1973.

Among the noteworthy features of this stochastic process, two properties, namely covering all distribution functions and being conjugate, are very applicable. This stochastic process is a noninformative prior. Besides, these features make it easier to work with. As such, due to no restrictions on parameters, these features make it more accurate to infer with.

The development of Bayesian computational methods, such as the MCMC method, has led to the use of the Dirichlet process as the prior distribution of nonparametric Bayesian problems, nonparametric regression, and nonparametric density estimation.

In Chapter 3, we first study the Dirichlet distribution. We then examine the Dirichlet process and some of its features. After examining this process and how it behaves for different parameters, in the second part of this chapter, we examine Polya's urn scheme and its relation to the Dirichlet process. In the third section, we introduce the predictive method using the Blackwell-MacQueen urn model, and finally, we examine one of the features of the Dirichlet process, namely data clustering using the Chinese restaurant process (CRP).

In Chapter 4, we first examine mixture models and then introduce a regression model called the "Semiparametric Linear Logarithmic Regression", which is used in accelerated lifetime tests. In the next section of this chapter, we briefly study the methods for determining the base distribution as well as the precision parameter of the Dirichlet process. We then express the Dirichlet process mixture model using a kernel, and finally, we present the simulation algorithm for estimating unknown parameters using the MCMC method.

In Chapter 5, we show the performance of the model in accelerated lifetime tests using the Dirichlet process model, using the Weibull kernel, and we will use two illustrative examples to analyze the results.

In order to see the goodness of fit of the estimated models, it is necessary to compare it with the empirical estimation of the data distribution, obtained nonparametrically without any constraints. Besides, to get more understanding of the merit of the nonparametric Bayesian method, another model is used to test the parametric accelerated lifetime whose parameters are estimated by the maximum likelihood method.

Since the empirical distribution function can be considered as a criterion to determine the goodness of fit, we first briefly review the empirical distribution function in Chapter 5 and then illustrate the general algorithm introduced in Chapter 4 in the particular mode. Finally, using two real examples, we present an analysis of the Dirichlet process mixture model, which is a nonparametric Bayesian model, and a parametric model.

Nezameddin Faghih
Ebrahim Bonyadi
Lida Sarreshtehdari

Acknowledgments

The authors would like to acknowledge the encouragement and support of their families, friends, and colleagues, and all the reviewers who undertook reviews in the double-blind peer review process. They are also sincerely grateful to Cindy Renee Carelli, the executive editor; Erin Harris, the senior editorial assistant; and the rest of their incredible team at CRC Press.

Authors

Nezameddin Faghih is the UNESCO chair professor emeritus and the founding editor-in-chief of the *Journal of Global Entrepreneurship Research* (Springer). He has published more than 50 books and 100 research articles, and presented more than 120 invited talks in academia, industry, and professional meetings.

Ebrahim Bonyadi is an applied statistician in the areas of business and economics and is a researcher at the Global Entrepreneurship Monitor (GEM) Office of the Faculty of Entrepreneurship, University of Tehran. His scholarly research focuses on factors influencing entrepreneurship, business, and economic growth.

Lida Sarreshtehdari is an applied statistician focusing on entrepreneurship, with expertise in the Global Entrepreneurship Monitor (GEM) dataset. She is a researcher at the Global Entrepreneurship Monitor (GEM) Office of the Faculty of Entrepreneurship, University of Tehran. She has published several reports on domestic entrepreneurship since 2011.

1 Introduction

Business success requires the increasing value of the products and services provided by manufacturers. These products and services are welcomed by society when, in addition to ease of use and affordable prices, they have a high quality and reliability. In other words, guaranteeing a product for a long time will attract the consumer market.

One of the main concerns of manufacturers is how to increase the reliability of products. Therefore, efforts to use durable raw materials, the powerful production process, efficient design, and affordable prices are among the tasks performed by the area of quality management, engineering management, design engineering, and finally repair and maintenance.

The question here is: How do manufacturers guarantee a product? You may have heard that some manufacturers use fake warranties to attract consumers but, naturally, it doesn't take long for them to go bankrupt.

Determining a specific time period as a guarantee from the manufacturer to the customer requires careful and laboratory analysis of the product to model the approximate lifespan of that product and use the "average lifespan" index as a guarantee criterion for the customer.

The credibility of a product lies in its guarantee, reliability, and quality. Achieving product quality improvement methods requires the use of predictive methods in the field of probability and stochastic models. In other words, the study of product reliability by quality control managers is in close connection with statistical models that are often able to model the time-to-break and hazard rate of products (Berndt et al., 2001).

1.1 NEED FOR QUALITY

If we take a serious look at the history of industry and look closely at all human industries, we will honestly acknowledge how much the quality element has influenced the fate of productions and services. According to all industry claimants, it is the quality element that has played an effective and valuable role in the dynamics of manufactured products.

This suggests that the role of quality in all matters should be considered much higher and more important, because otherwise all sectors, whether producing the right product or service, will suffer a complete decline, and thus all the effort expended will be ineffective. It can be argued that quality, with its inherent characteristics, will lead to the acquisition of segments in a desirable and acceptable way, creating added value and increasing productivity and profitability. Besides, the survival of any organization requires the acceptance of the qualitative characteristics of its activities because the element of quality is an integral part of the life of the organization's workforces (Chittenden et al., 1998).

With proper management of affairs, human beings can be guided in the right wdirection for a desirable and efficient production, and this will not be possible except by creating material and spiritual motivations for the employees of the organization. By accepting the above formula, you can also access the key points of market presence. Since science and technology are constantly evolving, it is impossible to ignore this issue for success.

In a word, the key to success in such a turbulent market is a deep understanding of competitors and the target market. Competitors may have fully grasped the concepts of valuation and have entered the field of production and services with extensive studies and full knowledge of market trends. They know exactly where they are in the market, who is ahead, who is in the middle of the field, and who is behind the scenes (Al-Assaf and Schmele, 1993).

They often thoroughly study about what customers expect from the product and what the craftsmen and managers will do in the future, as well as they know the region where the customers have more demands. How can they creatively and wisely satisfy customers in addition to reducing rising costs? How can they deal effectively with their industrial competitors in addition to this satisfaction?

Undoubtedly, the most practical answer is to maintain quality at all stages of industrial and service activities. Whenever the issue of quality and its promotion is taken into account by all personnel, especially managing directors, it is to be hoped that the unlocked locks of the major sectors in the competitive market will be opened.

Market capture depends on a systematic and strong view of the element of quality in all its dimensions. Entering the slightest flaw in the coherence of this attitude will lead to another path for industry managers, which will be the result of gradual destruction and loss of market and customers. This is a warning to managers who view the element of quality and reliability as trivial and insignificant.

1.2　QUALITY MANAGEMENT

Quality management is a guarantee for the sustainability of products and services produced by factories and is one of the effective parts of the success of businesses. Although its background dates back to the beginning of history, the new concept of quality is rapidly developed in the 20th century. According to one of the concerns, quality management means monitoring the process of manufacturing and product production to ensure that the product matches what the designer or customer desires. This monitoring includes everything from receiving and ordering raw materials to after-sales services. Thus, it covers a wide range of activities.

One of the activities related to quality management is quality assurance and quality control (Faghih and Nobari, 2004). Quality management consists of four main areas: quality planning, quality assurance, quality control, and quality improvement. Today, quality management plays a key role not only in the field of production but also in improving organizational behavior.

What is needed for growth in the field of quality management and providing quality products and services is a strong ability to study the quality of products and prevent an increase in the rate of defective products (Ghahramani, 2003).

1.2.1 Quality Management Parameters

The ability to manage quality in different persons may be influenced by many factors. But what makes a significant difference in profitability, reducing defective products, increasing customer satisfaction, and business development is the use of the most efficient method of evaluating the production process from raw material to end-user satisfaction level (Kimberly and Minvielle, 2000).

Different criteria and constructions are used to measure and evaluate the performance of each organization. Some of the indicators that are more common than other performance appraisal criteria are as follows:

1. **Effectiveness** has a general meaning. Simply put, effectiveness can be called "doing the right thing". Many scientists define the effectiveness as the understanding and achieving goals by the organization, and it is said that other concepts such as flexibility, acquisition, growth of resources, and improvement of human resources have been proposed for it.

    In other words, effectiveness reflects how much effort has been put into achieving the desired results. In management literature, effectiveness is doing the right thing and efficiency is the right doing of things. The concept of effectiveness lies in the concept of efficiency. Efficiency and effectiveness are quantitative and qualitative measures, respectively.

2. **Efficiency** means the least time to do the most work with available resources. The growth of efficiency level is directly in the hands of managers. Increasing efficiency enhances productivity and helps achieve organizational goals. The term "efficiency" has a more limited meaning and is used in connection with activities within the organization. In other words, trying to be efficient means: What do we do for each input unit to get a more useful output?

    Organizational efficiency is a criterion of valid resources that are used to produce a unit of a product and can be calculated as a ratio of consumption to the product. If an organization can achieve a specific goal by spending fewer resources than another organization, it is said to be more efficient. In other words, efficiency is the ratio of the amount of work done to the amount of work that shall be done.

3. **Innovation** is manifested creativity which has been reached the stage of action. In other words, innovation is the realization of a creative idea or the introduction of a new product or service to the market. Innovation is the use of mental abilities to create a new thought or idea.

4. **Flexibility** or resiliency is generally the ability of an organization to understand environmental change and then respond quickly and efficiently to that change. This environmental change can be a technological change and in conformity with the customer's needs. The term "flexibility" describes the speed and power of response when confronted with internal and external events in an organization.

5. **Quality:** Although the standard quality is complex and multiple, throughout the history of the emergence of this branch of science, experts have

defined quality from different perspectives. According to Harvey and Green (1993), quality has both objective and subjective aspects. The objectivity of an object is the reality that created the object, and the mindset about an object is the desirability or value of its physical properties. These two concepts are closely related. Other definitions of quality are as follows:
- The degree of conformity of the product produced on the needs of customers (Juran, 2004);
- The quality is consistent with the needs that are expressed over design properties (Reeves and Bednar, 1994);
- Quality has a changeability aspect and can be applied according to customer needs (Abbott, 1956).

1.3 QUALITY DETERMINANTS

Quality management can be widely used in various sciences, including engineering, medicine, chemistry, botany, and so on. In particular, manufacturers often expect quality control experts to provide them with information about system performance (i.e., the maximum remaining time of system life) and the reliability of the products. In fact, in industrial and engineering fields, there are generally several concerns that can be responded by applying relevant statistical methods like survival and reliability analysis. Here are eight of the manufacturers' concerns for product quality management and control:

A. **Performance:** Is the product capable of working well?

Potential customers usually evaluate a product in terms of different functions and how each of these functions works. For example, a software can be evaluated to find out what it is capable of doing. By performing this evaluation, it may be determined that the software in question is far superior to other similar software in terms of processing speed.

As another example, consider an electronic piece inside a computer. A closer look at a collection of different types of this part may reveal that some parts that are cheaper have a longer lifespan. This analysis will not only increase knowledge of how to produce long-lasting parts but also prevent additional costs.

It should be noted that quality management does not solely mean spending a lot of money to increase the quality of that product. Sometimes it is possible to increase the quality several times by even changing the weather conditions of the factory (like cooling devices).

But really, how can we recognize the parameters affecting product quality? How can we carefully examine the lifespan of products that may last for decades? In order to examine the lifetime of products, should we wait for decades to see the product's failure?

B. **Reliability:** How often does the product spoil?

Complex products, such as most home appliances, cars, or airplanes, will need to be repaired during their lifetime. For example, one would expect a car needs to be repaired at times, but if it needs to be repaired continuously, then we can say that the car is unreliable. In the automotive industry, the customer's view of quality is significantly influenced by such dimensions.

C. **Durability:** How long does the product last?

Durability is the useful lifespan of a product. Obviously, consumers want products that perform satisfactorily over a relatively long period of time. For example, the automotive and home appliance industries are among the industries in which this quality dimension is considered important for most of its customers.

D. **Repair capability:** How easy is the repair of the product?

In the industry, customers attach great importance to the speed and cost of repairing their appliances. Examples include the automotive and home appliance industries.

E. **Beauty:** What does the product look like?

This dimension describes the quality of the product's appearance, taking into account factors such as shape, color, model, packaging method, tactile characteristics, and other similar properties. For example, beverage companies can be exemplified because beauty plays an important role in competing in the global market.

F. **Features:** What does the product do?

Customers often consider products that have different characteristics and are superior to competitors' products in this regard to be quality products. For example, a spreadsheet software package that has the ability to perform statistical analysis may be preferred over similar software that does not have this feature.

G. **Perceived quality:** What is the reputation of the product or company?

In most cases, a customer buys a product based on the organization's reputation for the quality of its previous products. This reputation is directly affected by the visible problems of the product. On the other hand, it is important to call for a major repair of the defective product by the manufacturer (after-sales service) and how to deal with the customer at the time of reporting. "Goodwill of producer" and "purchase of the product by the customer" are closely related.

For example, if a person uses a certain airline to do business and the flight is almost always scheduled and his/her equipment is never damaged, then this person prefers to always use this airline.

H. **Compliance with standards:** Has the product been designed exactly as intended?

We usually know products with quality that meets the predetermined requirements for it. For example, how much does the hood of a new car be assembled on the produced car? Do all the components match? If the product is large or small, the order of the system will be disrupted and this will cause the car to act inappropriately and inconsistently with the designer's opinion.

The traditional definition of quality is based on the view that products and services must meet the needs of their users. In fact, one key point to keep in mind when it comes to a product is that the product must meet the needs of the people who use it.

Among all determinant factors of product quality (i.e., performance, reliability, durability, repair capability, beauty, features, perceived quality, and compliance with predetermined standards), reliability may increase product quality assurance more than others.

From a statistical and reliability engineering point of view, a product that is reliable also has quality (Condra, 2001). Therefore, in most processes related to quality control, in the final stage, the reliability and/or durability of that product is also examined.

Reliability analysis of a product requires having remarkable data size from its various features (including longevity).

1.4 FACTORS AFFECTING RELIABILITY

The history of the use of reliability may trace back to the early 1930s, when the applied probability entered into issues related to the science of electronic engineering. With the passage of time, the term gradually entered other disciplines and applied statistical and probability equations in most sciences. For example, Nelson (1990), Meeker and Escobar (1998), Faghih and Hamedi (2007), Kapur and Lamberson (1977), Kececioglu (2002), Faghih (1989, 1998), and Faghih et al. (2013) worked on quality, reliability engineering, maintenance engineering, and engineering in design. Ferguson (1973), El-Aroui and Soler (1996), and Yuan et al. (2014) worked on nonparametric methods to develop reliability concepts. Chen (1994), Cai (2012), Faghih and Najafi (2004), Singer (1990), Faghih and Loghavi (2007), and Kai-Yuan et al. (1991) worked on the reliability methods using the fuzzy approach.

Quality, which is a concept derived from reliability, has always been of great interest to the consumer community throughout history. So far, billions of dollars have been spent on improving the quality and reliability of a product. But what is important is the growth and development of factories and manufacturing companies in order to improve product quality.

Expanding the application of statistical methods and probability equations in defining the concept of quality and reliability has also led to the creation of interdisciplinary branches. Engineering management, maintenance engineering, maintenance planning, quality control, quality management, project management, manufacturing, and industrial management are some of the disciplines that are closely related to the concepts of quality and reliability (Bell and Halperin, 1995; Faghih and Yuli, 2007).

In general, reliability can be defined as the ability of a system or subsystem to perform a specific and predefined mission correctly under certain conditions and over a period of time, usually expressed in a number of probabilistic parameters.

From the engineering point of view, reliability and its applications in systems may be discussed from different aspects as follows:

- the appropriateness of the system or device for a specific purpose until a certain time;
- the capacity of the system or device to perform the mission designed for it;
- system or device resistance to failure;

- the probability that the system/device at work will be able to perform the task well within a specified time frame;
- the probability of having low-cost repairs.

But the cases that make a product more reliable and increase its quality are things as design, production process, transportation, repair, and maintenance.

1. **Design:** One of the important issues in design is that the more elements and parts used, which are arranged in a linear order, the lower the reliability of a system. Therefore, how the elements communicate logically with each other is effective in ensuring the reliability of the whole set but, in general, it can be said that in the design, a long sequence of parts or subsystems should be avoided. When designing, the system should be inspected and parts and components that have a great impact on reliability should be identified in order to take measures for improving the reliability of the entire system. Also, since after-production and repair services are important parameters in the reliability of parts and systems that can be repaired, the design should be such that repair and after-sales service can be done easily.

    An issue regarding the after-sale services may refer to maintenance planning and scheduling that may be demanded by purchasers. Optimal, minimal, and minimax repairs are the options presented by product sellers, which have been formulated by statistical methods.

    Maintenance planning studies by applying parametric methods, non-parametric methods, fuzzy control, and classical statistics have been done by many scholars such as Palmer (2006), Khan and Haddara (2003), Faghih and Zadeh (2010), and Cassady and Kutanoglu (2005).
2. **Production:** In the production process, quality control techniques must be used to minimize the risk of production and the disadvantages that occur concurrently with the production process. Especially during production, more attention should be paid to elements that have lower reliability or are more sensitive so as not to fail during assembling or production.
3. **Transporting:** At the time of use, reliability strongly depends on how the product was shipped and how it was handled during transportation. Therefore, good and proper packaging is one of the effective factors in maintaining the product during transportation and as a result, product reliability during consumption.
4. **Repair and maintenance:** Repair and maintenance play an important role in maintaining and restoring reliability. System reliability is directly related to maintenance and repair conditions.

Following the probability models derived from the classical, fuzzy, and Bayesian statistical approaches will lead to greater success in quality control, maintenance engineering, design engineering, survival analysis, and management engineering. Therefore, the need to become familiar with the methods of studying the reliability of products has led to the publication of many scientific articles and books that are being developed every day. One of the most complete methods of reliability analysis is to use an unlimited approach that can be generalized to many variations. The nonparametric Bayes has acceptable accuracy without any restrictive assumptions, despite many difficulties.

What will be discussed in this book is not the methods of quality control management, design, repair and maintenance, and product quality improvement, but the presentation of a comprehensive model for modeling product lifespan to study its reliability over time.

Hence, in the upcoming chapters, we will take a step-by-step process of using the most powerful predictor method in analyzing lifetime data namely the nonparametric Bayes method with mixed models.

2 Quality and Reliability

In recent years, the application of statistical methods in various scientific branches has led the field to be considered as a powerful tool in the development of science and technology.

Today, it can be argued that the use of statistical techniques in the analysis and recognition of various phenomena is unavoidable. The use of statistics science in modeling and analyzing stochastic and uncertain phenomena, especially in different branches of engineering and industry, is one of the most important areas for the application of this science. The advancement of technology and the rise in the quality of life of people through industrial achievements have created a tough competition between manufacturers' ideas, products, and services of higher quality.

In the production of strategic products, if a company fails to deliver the quality and capability of its products of the desirable standards, it had to soon withdraw from the competition. Rapid advances in technology, the development of modern products/services, the intensification of global competition, and the growing expectations of customers have brought new pressures on high-quality product makers.

Customers expect to buy products that are reliable and with a long lifespan. Systems, machines, and devices should be able to operate at high health under standard conditions for a specified period.

Reliability is usually defined as the probability that a system, machine, or device will work well under normal circumstances for a given time. Improving reliability is an important part of the overall impression of product quality improvement. There are many definitions of quality, but the general consensus is that "an unreliable product is not a quality product". According to Kromholtz and Condra (1993) and Condra (2001), reliability is quality over time.

Reliability improvement programs for the creation of products require numerical methods to predict and evaluate different aspects of the reliability of a product. In many cases, this vision involves collecting reliability data from studies such as experimental tests or accurate monitoring of products' failure time.

In this chapter, we first discuss some of the features of reliability data. Besides, since we need some functions to interpret the characteristics of the survival data, we introduce some useful functions briefly.

In the subsequent section, we refer to censored data and accelerated lifetime tests (ALTs), which are one of the most important discussions in the analysis of reliability data, and also introduce several ALT models. Since the data used in this book is experimentally accelerated by an accelerator factor, we also describe the relationship between normal lifetime and stress levels. Finally, according to this issue that we will use Bayesian simulation methods (like the Markov chain Monte Carlo (MCMC), slice sampling, and importance sampling), hence, all methods will be tried to be described throughout this book.

2.1 SOME REMARKABLE PROPERTIES OF SURVIVAL DATA

Reliability (also called survival) dataset may contain one of the following features. Analyzing each one of these features entails applying a particular statistical method.

a. Reliability data are usually censored, i.e., the exact time of failure is not known. The most important reason for censorship can be to analyze the lifetime data before all units fail. In general, censor comes from the fact that actual response values (such as failure times) cannot be observed for some or all of the study units. Therefore, censored observations create a boundary or boundaries on real failure times.
b. Many reliability data are modeled using distributions. For instance, for positive random variables, Exponential, Weibull, Gamma, and Log-normal distributions are considerable. Since the domain of normal distribution can also be negative, in practice in a small number of cases, this distribution is used as a model of the lifetime of a product.
c. Most arguments and predictions in the area of reliability require extrapolation. For example, you might want to estimate the ratio of units that do not work after 5000 hours, while the experiment lasts for up to 1000 hours (which is called "time extrapolation"). You might also want to estimate the time-to-break of 1% of products that work at 50°C, while the test is carried out at 85°C (which is called "operation extrapolation").
d. Normally, in reliability analysis, traditional models, such as average and standard deviation, are not at the top of the list. Design engineers, managers, and customers are interested in certain criteria of product reliability or features of the distribution of failure time (such as failure probabilities, lifetime distributions, and failure rates).
e. The fitting of the model, especially for censored data, requires the use of numerical methods by the computer, and there is often no accurate theory for statistical inferences.

As is coming out of the characteristics of reliability data, a large part of the collected reliability data include information about the failure time of some components or the entire system. Therefore, the criterion which is widely used for analyzing the reliability of a product is the distribution of the failure time. Note that the upper case letter T will be used to represent a continuous and nonnegative random variable that describes the failure time of a unit or system.

2.2 IMPORTANT FUNCTIONS FOR ASSESSING FAILURE TIME

Features of the failure time, T, can be assessed using the functions that represent the relevant properties. Some of these functions are cumulative distribution function (CDF), probability density function (PDF), survival function (also known as reliability function), and risk or hazard function (HF).

2.2.1 Cumulative Distribution Function

The CDF of the random variable T is noted by $F(t)$. This function is mathematically defined as follows:

$$F(t) = P(T \leq t), \quad t \geq 0 \tag{2.1}$$

The CDF applies to the following three properties:

a. $F(t)$ is a non-descending function;
b. $F(t)$ is a continuous function from the right;
c. $F(\infty) = 1$ and $F(0) = 0$

Note that $F(t)$ can be interpreted as the ratio of units of a population that their lifespan does not exceed t.

2.2.2 Probability Density Function

The PDF of the continuous random variable T is defined as the derivative of $F(t)$ with respect to t. That means if $f(t)$ is the density function of variable T, then

$$f(t) = \frac{dF(t)}{dt} \tag{2.2}$$

Obviously, $f(t)$ is true in the two following conditions. In fact, any function with the two following conditions can be considered as a PDF.

a. for $t \geq 0$, $f(t) \geq 0$
b. $\int_0^{+\infty} f(t) dt = 1$

The PDF can also refer to the relative frequency of failure times. Although the PDF is less important than other functions in reliability applications, it is widely used in the development of technical results. But working on the PDF will result in discovering the useful methods to study reliability variables.

2.2.3 Survival Function

Each statement, description, and interpretation as regards the lifetime variable T is extracted from $f(t)$ and $F(t)$. However, the fundamental function in the reliability issues whose duty is to determine more accurate characteristics of variable T is the survival function (also known as the reliability function). The reliability function of the random variable T is defined as follows:

$$R(t) = \int_t^{+\infty} f(x) dx = P(T \geq t) = 1 - F(t); \quad t \geq 0 \tag{2.3}$$

According to this definition, it is obvious that $R(0) = 1$. This means the reliability of a system at the time 0 is 1. As well as $R(+\infty) = 0$; this means with the passage of time the reliability of the system will be declined to zero. Also, $R(t)$ is a nonincreasing function, which means that if $0 < t_1 < t_2$, then $R(t_1) \geq R(t_2)$. That is, the reliability of the system at the time t_1 is not less than reliability at the time t_2.

2.2.4 Hazard Function

The HF or hazard rate function is one of the most important statistical tools in reliability studies and longevity issues.

Consider the random variable T with the CDF $F(t)$ and the PDF $f(t)$. The basic question that comes up here is: "if we assume that the system is still working well at time t, what is the probability of immediately breaking down of this system at the time t?"

In other words, for the small amount of δ, we are interested in finding the answer to the following conditional probability:

$$P(t < T < t+\delta | T \rangle t) = \frac{P(t < T < t+\delta)}{P(T > t)}; \quad t \geq 0, \delta > 0 \qquad (2.4)$$

Clearly, $P(T > t)$. If the above conditional probability is divided by δ and its limit be calculated when $\delta \to 0$, then it will result in a function called the HF.

If the HF of the continuous random variable T gets denoted by $h(t)$, then it can be calculated as follows:

$$h(t) = \lim_{\delta \to 0} \frac{P(t < T < t+\delta)}{\delta R(t)} = \lim_{\delta \to 0} \frac{R(t) - R(t+\delta)}{\delta R(t)} = \frac{f(t)}{R(t)}; \quad t \geq 0 \qquad (2.5)$$

The last equivalency results from the fact that

$$f(t) = \lim_{\delta \to 0} \frac{R(t) - R(t+\delta)}{\delta} \qquad (2.6)$$

The HF (also known as failure rate function in the reliability analysis) plays a crucial role in most survival analyzes. It should be noted that the HF is not the probability of the system failure (i.e., it does not necessarily take between zero and one). Additionally, this quantity represents the rate of degradation of the system over time and can measure the failure rate of the system.

This function can appear in five states including constant failure rate, increasing failure rate, descending failure rate, unimodal, and U-shape. The last one, as shown in Figure 2.1, has been widely used in engineering reliability and can be applied for the interpretation of varied phenomena.

As shown in Figure 2.1, the system's failure function is decreasing for the initial period. In reliability analysis, this period is known to be the early life period (or, equivalently, the burn-in period).

Quality and Reliability

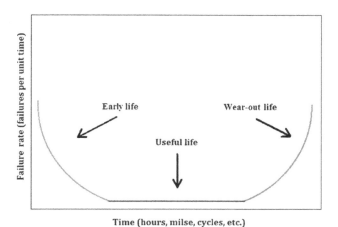

FIGURE 2.1 U-shape hazard rate. (Authors' own figure.)

During the next time interval, the hazard rate is almost constant, which is known as the system's useful life. Finally, in the third period, the hazard rate begins to increase, which is called the wear-out life.

If $h(t)$ is an increasing function of t, it reflects that the system will get degraded and its reliability will decrease as time passes. Moreover, if $h(t)$ is descending, then the system will face a negative degradation, meaning that its reliability increases over time.

To get more understanding, you can refer to a car's lifespan period. At the beginning of lifespan, it has the highest probability of breaking down and also may face early deterioration during this period. Due to lack of a proper join of parts, the first period of the lifetime may hold the most probability of failure.

Example 2.2.1

For more details, suppose random variable T comes from the Weibull distribution with shape parameter $\alpha > 0$ and scale parameter $\lambda > 0$.

The PDF, CDF, reliability function, and HF of the Weibull distribution are respectively achievable as follows:

$$\begin{aligned} f_T(t;\alpha,\lambda) &= \alpha\lambda t^{\alpha-1} e^{-\lambda t^\alpha}, \\ F_T(t;\alpha,\lambda) &= 1 - e^{-\lambda t^\alpha}, \\ R_T(t;\alpha,\lambda) &= e^{-\lambda t^\alpha}, \\ h(t;\alpha,\lambda) &= \alpha\lambda t^{\alpha-1}. \end{aligned} \quad (2.7)$$

where $t \geq 0$, $\alpha > 0$, $\lambda > 0$.

Figure 2.2 illustrates the simulated curves concerning these four functions when the shape parameter is equal to 2 and the scale parameter is 0.9. To use programming codes of this figure, see Section B.1 in the Appendix.

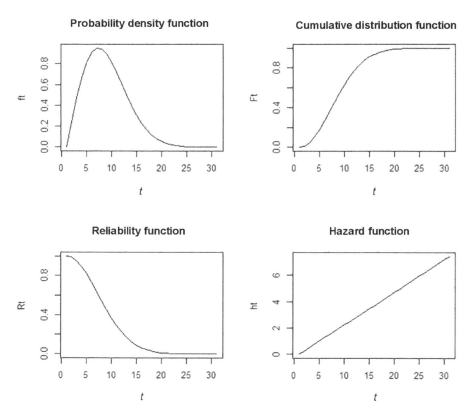

FIGURE 2.2 The simulated probability density function, cumulative distribution function, reliability function, and hazard function of the Weibull distribution when the shape parameter is equal to 2 and the scale parameter is equal to 0.9. (Authors' own figure.)

Because the HF is strongly influenced by the shape parameter, this function will appear in different trends including increasing, decreasing, and constant modes. To observe the impact of shape parameter on the behavior of HFs, see Figure 2.3. To use programming codes of this figure, see Section B.2 in the Appendix.

2.2.5 Quantile Function

Quantile t_p is the inverse of the CDF and refers to the time that a certain ratio (like p) of the population fails to work well. That means

$$t_p = F^{-1}(p) \tag{2.8}$$

In general, for $0 < p < 1$, p^{th} quantile of distributions $F(t)$ is the smallest value when the following condition holds:

$$P(T \leq t) = F(t) \geq p = P(T \leq t_p) = F(t_p) \tag{2.9}$$

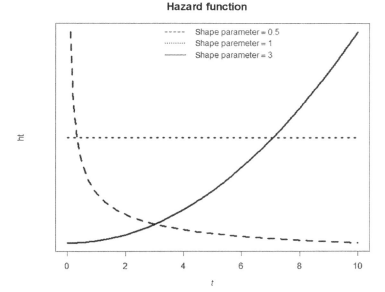

FIGURE 2.3 Hazard function of the Weibull distribution when the scale parameter is equal to 0.9 while the shape parameter is changing. (Authors' own figure.)

2.3 CENSOR

The accurate analysis of lifetime data requires basic information on product lifespan. Since technological advancement has increased the quality of the products, in order to obtain information related to the lifetime of the products, there is a great need for time and cost, which in some cases (including product cost) may be opposed by managers and investors. One way which is being used in such cases is to provide ideas for a reduction in tests' period of time. This idea, in addition to saving time, also reduces costs.

Hence, to determine the approximate time of a product's failure time, the use of data censoring methods is a clever and affordable method, albeit is associated with the hardships and difficulties. Various methods are being used to censor the lifespan of products in the industry, but each way has its place.

Some of the most frequency data censorship methods are listed below:

- Left Censoring
- Right Censoring
- Random Censoring
- Progressive Censoring
- Hybrid Censoring
- Interval Censoring
- Double Censoring
- Type I Censoring
- Type II Censoring

For more details, the books written by Nelson (1990) and Meeker and Escobar (1998) can be studied.

Since the "right censoring type one" data will be used throughout the present book, we briefly describe the structure of this type of censorship.

Suppose we inspect n pieces of a product at time $t = 0$ and then will be waited until time C to see complete failure times or incomplete data. Suppose r ($r = 0,1,2,..., n$) is the number of products that have been failed before time C and their exact time of failure is observed. The only information about the $n-r$ products is that their life-time is more than C. In this situation, it is said that $n-r$ products have been censored. Note that, this type of censoring is known as the type I censoring method.

In type I censoring, if $t_1, t_2, t_3, ..., t_r$ are the accurate life-time of the r products that have failed before time C as well as if remaining $n-r$ products are censored, the likelihood function of these n products is as follows:

$$L(\theta) = \prod_{i=1}^{r} f(t_i, \theta) \prod_{j=r+1}^{n} R(C, \theta) = \prod_{i=1}^{r} f(t_i, \theta) R(C, \theta)^{n-r} \qquad (2.10)$$

where $R(C, \theta)$ is the reliability function corresponding to $f(t)$ at time C. The logarithm of the likelihood function can be rewritten as follows:

$$\iota(\theta) = \sum_{i=1}^{r} \ln\left(f(t_i, \theta)\right) + (n-r) \ln R(C, \theta) \qquad (2.11)$$

Since it is not known how many failures will be observed before the constant time C, r can be considered as a random variable. Additionally, this variable can be $0, 1, ..., n-1$ or n.

Accordingly, it can be argued that r is a random variable that comes from a binomial distribution with parameters n and $F(C, \theta)$ in which $F(C, \theta)$ denotes the CDF relevant to the $f(t, \theta)$. Using the above likelihood function, the maximum likelihood estimator of the parameter θ can be obtained.

2.4 ACCELERATED LIFETIME TESTS

Today, providing products with better characteristics and higher reliability for manufacturers is a major issue. Consumers expect products to have high quality and work well for a long time.

Therefore, quality and/or reliability guarantee is considered as one of the most important features of the products in view of the customers' community. A more accurate estimate of reliability through testing the products at the various steps of product production is a fundamental act of quality control that lead to quality improvement.

Since many modern products are designed to be used without fail for decades or more, testing these products in natural conditions will take a very long time and will naturally be costly. Hence, in order to accelerate the lifetime of products to reduce the time and cost of testing, many methods have been developed by scholars. One of the most accurate and applicable methods is known as the "Accelerated Lifetime Test (ALT)".

In other words, ALTs are used to get faster information on the lifetime distribution of a product. ALTs are widely used in manufacturing industries, especially for accessing information about component reliability.

Generally, the tested units are under a stress level with more severe conditions than normal state that will result in shorter unit lifetimes than normal. Note that the accelerated data will be extrapolated to be converted to normal values.

The accelerator variable that affects the units' lifetime can be one such thing as temperature, voltage, electricity flow, mechanical load, humidity, vibration, and pressure. Depending on the product and test type, one of the accelerator factors is being used in ALTs. It should be noted that the accelerator variable is usually referred to as stress.

2.4.1 Accelerated Lifetime Tests Models

In general, accelerated tests are divided into two categories including ALTs and accelerated degradation tests. In some lifetime tests, none of the tested units may fail over time. Therefore, in such circumstances, the researcher will not have any information on the lifespan of the units under study.

But if the tested units are surveyed based on studies on degradation concept, it can be observed that, notwithstanding the units have not broken, each of them has lost some of their ability during the degradation process as well as it is likely that some units have progressed to the brink of failure.

Therefore, the use of accelerated degradation test methods will be useful. Since our goal in this study is to scrutinize the ALTs, in the following, we will describe the ALT models specifications.

In most ALTs, the stress on the unit under test is often as constant-in-time stress, step stress, and progressive stress. The first case is the method, which is used in this study, has a simple function, which means that the stress level for the unit or system under investigation is constant over time.

Step stress is applied to the test the units in the manner of stepwise in different periods of time. During each period, the stress level changes to a higher level.

In the progressive stress process, for the examination of the unit or system under view, the stress level increases constantly and steadily from the start time of the test to the occurrence of failure.

In ALTs, researchers shall take two important principles into consideration:

- First, the failure-time of units at a different stress level comes from a family of distributions;
- Second, a functional relationship between stress level and model parameters should be determined.

There are wide-ranging sources comprising an in-depth investigation of the accelerator functions. Mann et al. (1974), Nelson (1990), and Meeker and Escobar (1998) are the early scholars who have comprehensively discussed a lot of practical accelerator functions in their studies.

2.4.2 LIFETIME-STRESS RELATIONSHIP

The relationship between lifetime and stress level is often determined using a function. This function, that shows the relationship between the lifetime and the accelerator parameter, also includes a number of unknown constant coefficients, too.

Some of the most commonly used accelerator functions are as follows:

1. Power Law Relationship: $\vartheta_\theta = \gamma L^\theta$
2. Arrhenius Relationship: $\vartheta_\theta = \gamma e^{\frac{\theta}{L}}$
3. Inverse Logarithm Relationship: $\vartheta_\theta = \gamma (\log L)^\theta$
4. Exponential Model: $\vartheta_\theta = \gamma e^{\theta L}$
5. Inverse Linear Model: $\vartheta_\theta = \gamma + \theta L$

Where in the above functions, ϑ_θ represents the model. Further, L shows the accelerator variable or the stress level, and γ and θ are the constant unknown parameters of the model. Except for the Inverse Linear Model, all of the above models can be rewritten with a simple transformation on the accelerator variable into the form of the Power Law Model.

2.5 BAYESIAN APPROACH

Suppose θ is the ratio of the number of broken lamps to the number of lamps that a company produces. Since it is natural that owing to the degradation of factory machines, θ gradually increases and cannot always be considered as a constant value. So, the parameter θ can be considered as a random variable.

Hence, solving the statistical problems and issues with such visions (i.e., seeing any unknown parameter as a random variable) is discussed in the field of the Bayesian approach. All visions in the Bayesian approach are rooted in Bayes' law.

Given the nature of the Bayes principle, in the Bayesian method of statistical inference, the parameter θ is an unknown value that comes from the family of distributions. In other words, θ is considered as a random variable like W, whose possible values are in the parameter space Θ.

In addition, this variable has the CDF $G(\theta)$ and the probability function (also called the PDF) $g(\theta)$, and $g(\theta)$ is referred to as the prior distribution or the prior density function.

In fact, the prior distribution represents a widespread use of statistics in the context of the summary of prior information and knowledge about which parts of Θ are most likely to occur. In other words, before observing and collecting any data, the previous evidence of the data (prior density function) informs the researcher of the chance of θ in the probability space Θ. Also, through math and probability inferences, the researcher will be convinced that considering the prior information as a decisive factor is a completely logical argument, whilst ignoring the previous information will raise a big mistake. Therefore, a group of statisticians, known as Bayesians, declare that previous beliefs and prior information can be summarized in a distribution function form.

Quality and Reliability

The issue of prior distributions from the viewpoint of statisticians is generally controversial and generally refers to the attitude of individuals about probability. This has led to a dichotomy between classical statisticians and Bayesian statisticians. Notwithstanding the deep differences, a great deal of researches have been published by considering both classic and Bayesian methods. As well as, sharing essential insights into different methods which have been made by scholars in the classical or Bayesian approach, and also the composition of both branches, has developed the statistical methods.

According to explanations above, in this case, the Bayesian risk function in choosing δ, as an estimator relative to the prior distribution function G and the loss function L, which is notated by $r(G, \delta)$, is equal to

$$r(G, \delta) = E\{E\{L(W, \delta(X))\}\}$$
$$= E[R(W, \delta)] \qquad (2.12)$$
$$= \int_{\Theta} R(\theta, \delta) \, dG(\theta)$$

where $R(\theta, \delta)$ is named Risk Function. Further, we seek δ_B to minimize the value of the risk function. In this case, we refer to the estimator $\delta_B(X)$ with the title of the Bayesian estimator related to the prior distribution G under the loss function L.

According to the explanations above, we assume that the probability model is dependent on the parameter θ which is coming from the distribution family $\{F_\theta; \theta \in \Theta\}$ and probability density family $\{f_\theta; \theta \in \Theta\}$. As well as suppose θ is a known and observed amount of variable W with known distribution function $G(\theta)$ and density function $g(\theta)$ that we name it the prior distribution of variable W.

In this case, $f_\theta(.)$ is considered as a conditional density function of the variable X given $W = \theta$ in which the joint density function of X and W is obtained from the following equation:

$$f_{X,W}(x, \theta) = f_{X|W=\theta}(x, \theta) g_W(\theta) \qquad (2.13)$$

The conditional distribution of the W on the condition $X = x$ is assessable, which is known as the posterior distribution of the variable W.

The posterior PDF of the variable W is calculable through the above equation and the Bayes formula. See the following equation:

$$g_{W|X=x}(\theta) = \frac{f_{X,W}(x, \theta)}{f_X(x)} \qquad (2.14)$$

in which $f_{X,W}(.,.)$ represents the joint density function of variables X and W as well as $f_X(x)$ is the marginal distribution of variable X, and this marginal distribution can be obtained as follows:

$$f_X(x) = \int_\Theta f_\theta(x) \, dG(\theta) = \begin{cases} \sum_\theta f_{X,W}(x, \theta); & \text{if } W \text{ is discrete} \\ \int_\Theta f_{X,W}(x, \theta) \, d\theta; & \text{if } W \text{ is continuous} \end{cases} \quad (2.15)$$

Therefore, the posterior density function of the parametric variable W can be calculated as follows:

$$g_{W|X=x}(\theta) = g(\theta|x) = \frac{f_\theta(x) g(\theta)}{\int_\Theta f_\theta(x) \, dG(\theta)} \quad (2.16)$$

In many cases, we may be interested in using the information of θ from other sources, too. If the collection of all this information on θ is accessible in the form of a probability distribution function (i.e., if θ is considered as the value of a random variable), as remarked, the characteristics of θ can be obtained by applying the Bayes theorem and a mixture of present information and prior evidence.

Hence, the density function of the prior information is called the prior density, and the density function of the mixed information of both prior and present evidence is named the posterior density function. Note that the characteristic of "being a prior distribution" is not exclusively related to the distribution, but its assignment returns to the relation between distribution and x. For this purpose, traditionally, $G(\theta)$ is considered as the distribution of W before x (the prior distribution) and $G(\theta|x)$ is considered as the distribution of W after x (posterior distribution). Thus, if x_1 and x_2 are two units of the given data sequence, then the posterior distribution after x_1 is considered as the prior distribution before x_2.

Note that the following equation

$$f_X(x) = \int_\Theta f_\theta(x) \, dG(\theta) \quad (2.17)$$

is not affected by θ. Therefore, in calculating the posterior density, it is enough to consider $g(\theta|x)$ proportional to the product of $f_\theta(x)$ and $g(\theta)$ and then normalize it.

Since the main purpose in Bayesian inference is obtaining an approximation of the above integral with the use of data, the functions within the integral may not have a closed form in most cases. Hence, to approximate such integrals, various simulation methods are applied. The MCMC technique is one of the most common manners which can be expressed as follows.

2.6 MARKOV CHAIN MONTE CARLO METHOD

2.6.1 MONTE CARLO APPROACH

The Monte Carlo method is an approximate solution to math, statistics, and physics problems.

In fact, any method that attempts to solve a problem by generating random numbers is known as the Monte Carlo method. The Monte Carlo method makes it possible to simulate each parameter generated by a random process. But this is not the only usage of the Monte Carlo method. In many mathematical problems in which chance does not play any role, the probability models can artificially be determined by applying vast evidence. For this reason, the Monte Carlo method can be considered as a general solution to mathematical problems.

Since one of the major concerns in the Bayesian problems is the solution of the integral to find the posterior distribution, especially when the target function is a high-dimensional function, many methods for calculating these integrals have been provided by scientists such as Smith (1991), Evans and Swartz (1995), and Tanner (1996).

2.6.1.1 Monte Carlo Integration

The original Monte Carlo method was initially developed by physicists to compute complex integrals using random numbers, summarized as follows. Suppose our goal is to calculate a complex integral with a general form as follows:

$$\int_a^b h(x) dx \tag{2.18}$$

The idea of the Monte Carlo method for solving such integrals is dependent on the decomposition of function $h(x)$. In this approach, the function $h(x)$ is decomposed into two functions, namely, $f(x)$ and a PDF $p(x)$. $p(x)$ is defined on the interval (a, b) which is easy to be sampling.

With all these explanations, the above integral can be rewritten as follows:

$$\int_a^b h(x) dx = \int_a^b f(x) p(x) dx = E_{p(x)}[f(X)] \tag{2.19}$$

According to the above integral, it is shown that the answer to this integral can be obtained by calculating the mathematical expectation of variable $f(X)$, with respect to the density function $p(x)$.

Now, if $x_1, x_2, \ldots, x_n$ is an Independent and Identically Distributed (IID) sample generated from density function $p(x)$, then $f(x_1), f(x_2), \ldots, f(x_n)$ will be IID and are gained from the same density function.

By generalizing this argument, it would be derived that for any $i = 1, 2, \ldots, n$, $E_{p(x)}[f(x_i)]$ is an approximation of the integral.

According to **strong law of large numbers**, for a big enough sample, the above integral can be rewritten as follows:

$$\int_a^b h(x) dx = E_{p(x)}[f(X)] \cong \frac{1}{n} \sum_{i=1}^n f(x_i) \tag{2.20}$$

It should be noted that the above integral is a special state of multivariable functions. That is, if the function h is an n-variable function, then the calculation of the multivariable function integral is carried out by generalizing the previous one-variable argument as follows:

$$\int_a^b \int_a^b \ldots \int_a^b h(x_1, x_2, \ldots, x_n) dx_1 dx_2 \ldots dx_n \qquad (2.21)$$

After decomposing $h(x)$ into two functions $f(x)$ and $p(x)$, in which $x = (x_1, x_2, \ldots, x_n)$, the foregoing integral will be approximated by the mathematical expectation of function $f(X)$ with respect to PDF $p(x)$.

As a result, if there are k sets of samples with size n, which have been extracted from $f(x_1), f(x_2), \ldots, f(x_n)$, then given the strong law of large numbers, the considered integral can be written as follows:

$$\int_a^b \int_a^b \ldots \int_a^b f(x_1, x_2, \ldots, x_n) p(x_1, x_2, \ldots, x_n) dx_1 dx_2 \ldots dx_n$$

$$= E_{p(x)}[f(X)] \cong \frac{1}{k} \sum_{i=1}^k f(x_{i1}, x_{i2}, \ldots, x_{in}) \qquad (2.22)$$

The mentioned method that is capable to estimate complex integrals is called the Monte Carlo Integration Method. This method is also widely used to estimate posterior distribution or marginal posterior distributions in Bayesian analysis.

To estimate the following integral,

$$I(y) = \int f(y|x) p(x) dx \qquad (2.23)$$

It is also sufficient to create a large-enough sample of the density function $p(x)$ and, finally, apply the Monte Carlo method with the use of the following equation:

$$\hat{I}(y) = \sum_{i=1}^n f(y|x_i) \qquad (2.24)$$

The standard error of this estimate can be approximated by

$$ES^2(\hat{I}(y)) = \frac{1}{n}\left(\frac{1}{n-1}\sum_{i=1}^n \left(f(y|x_i) - \widehat{I(y)}\right)^2\right) \qquad (2.25)$$

This equation shows the large sample will derive a more exact estimation.

2.6.1.2 Importance Sampling

In line with the calculation of the complex integrals, the decomposition of function $h(x)$ may be hard to do, as well as for whatever reason, it might the sample generation of the specified PDF with the desired size be impossible. The method with wide

Quality and Reliability

applicability, which is implementable in such circumstances, is referred to as importance sampling.

The enforcement procedure for this method is as follows:

Suppose PDFs $p(x)$ and $q(x)$ have the same domains. In this case,

$$\int f(x)q(x)dx = \int f(x)\left(\frac{q(x)}{p(x)}\right)p(x)dx = E_{p(x)}\left[f(X)\left(\frac{q(X)}{p(X)}\right)\right] \quad (2.26)$$

Therefore, in order to calculate this integral, it is sufficient to calculate the following mathematical expectation by applying the Monte Carlo method when the target density function is $p(x)$:

$$\int f(x)q(x)dx \cong \frac{1}{n}\sum_{i=1}^{n}f(x_i)\left(\frac{q(x_i)}{p(x_i)}\right) \quad (2.27)$$

It is worth mentioning that in line with estimating this integral using the Monte Carlo method, there exist other methods to gain samples such as the rejection and acceptance method, the inverse distribution function method, and the variable change. After having a big sample size of the considered PDF, numerical calculation of the integral using the method of Monte Carlo would be easy to do.

2.6.2 MARKOV CHAIN

2.6.2.1 Definitions

Definition 2.6.1

The sequence of random variables $\{X^{(t)}, t \geq 0\}$, which all $X^{(t)}$ are in the countable set S, is called "Homogeneous Markov Chain", with the state space S when the following is true:

$$P\{X^{(t+1)} = j | X^{(t)} = i, X^{(t-1)} = i_{(t-1)}, \ldots, X^{(0)} = i_{(0)}\}$$
$$= P\{X^{(t+1)} = j | X^{(t)} = i\} = p_{ij} \quad (2.28)$$

p_{ij} is named the one-step transition probability from state i to state j and is defined as

$$P = [p_{ij}]_{i,j \in S} \quad (2.29)$$

In this case, P is defined as the one-step transition probability matrix for chain $\{X^{(t)}, t \geq 0\}$. Clearly, for any $i, j \in S$, $p_{ij} \geq 0$ and $\sum_{j \in S} p_{ij} = 1$.

Similarly, if we define

$$P^{(m)} = \left[p_{ij}^{(m)} \right]_{i,j \in S}$$

$$P_{ij}^{(m)} = Pr\{X^{(t+m)} = j | X^{(t)} = i\}$$

(2.30)

Then, $p_{ij}^{(m)}$ is the m-step transition probability and $P_{ij}^{(m)}$ is the m-step transition probability matrix of chain $\{X^{(t)}, t \geq 0\}$.

Obtaining the amount of probability that the chain at the instant t is in position j is one of the important issues that gives the researcher comprehensive information about the chain, and its mathematical notation is written as follows:

$$\Pi_j^{(t)} = \Pr(X^{(t)} = j)$$

(2.31)

For calculating this probability, the Chapman–Kolmogorov equation can be useful. For example, to calculate the probability that a chain at time $t+1$ is at position i can be achieved as follows:

$$\Pi_i^{(t+1)} = \Pr(X^{(t+1)} = i) = \sum_k \Pr(X^{(t+1)} = i | X^{(t)} = k) \Pr(X^{(t)} = k)$$

$$= \sum_k p_{ki} \Pi_k^{(t)}$$

(2.32)

For as much as

$$p_{ij}^{(2)} = \Pr\{X^{(2)} = j | X^{(0)} = i\}$$

$$= \sum_{k \in S} \Pr\{X^{(2)} = j, X^{(1)} = k | X^{(0)} = i\}$$

$$= \sum_{k \in S} \Pr\{X^{(2)} = j | X^{(1)} = k\} \Pr\{X^{(1)} = k | X^{(0)} = i\} = \sum_{k \in S} p_{ik} p_{kj}$$

(2.33)

Therefore, it can be said that $P \times P = P^2$. By extending this argument to the m-step, the Chapman–Kolmogorov equations can be achieved. That is,

$$P^{(m)} = \overbrace{P \times P \times \ldots \times P}^{m}$$

$$P^{(m)} \times P^{(l)} = P^{(m+l)}$$

(2.34)

These equations state that with a one-stage transition probability matrix, the multistage transition probability matrix can be obtained. Therefore, the one-stage transition probability matrix P completely characterizes the Markov chain structure. In other words, each Markov chain can be expressed by the one-stage transition probability matrix.

2.6.2.2 Chain Structure

The initial status of a chain is represented by $X^{(0)}$, and after obtaining the distribution of $X^{(0)}$, the distribution of other statuses ($X^{(t)}$) will be achievable using the following method.

Suppose $X^{(0)}$ and $X^{(1)}$ have the distributions $\Pi^{(0)}$ and $\Pi^{(1)}$, respectively. For simplicity, we show (not so arbitrarily) these distributions as follows:

$$\pi_i^{(0)} = Pr\left(X^{(0)} = i\right),$$

$$\Pi^{(0)} = \left[\pi_i^{(0)}\right]_{i \in S}, \quad (2.35)$$

$$\Pi^{(1)} = \left[\pi_j^{(1)}\right]_{j \in S}$$

Using the Chapman-Kolmogorov equation,

$$\pi_j^{(1)} = Pr\left(X^{(1)} = j\right) = \sum_{i \in S} Pr\left\{X^{(1)} = j | X^{(0)} = i\right\} Pr\{X^{(0)} = i\}$$

$$= \sum_{i \in S} \pi_i^{(0)} p_{ij} \quad (2.36)$$

Now, according to matrix symbolizing,

$$\left(\Pi^{(1)}\right)^T = \left(\Pi^{(0)}\right)^T \times P \quad (2.37)$$

Similarly, if the distribution of $X^{(t)}$ is shown with $\Pi^{(t)}$, then

$$\left(\Pi^{(t)}\right)^T = \left(\Pi^{(0)}\right)^T \times P^{(t)} \quad (2.38)$$

This shows that the distribution of $X^{(t)}$, for any $t \geq 0$, only depends on the matrix P and initial distribution $\Pi^{(0)}$. In another word, P and $\Pi^{(0)}$ are able to explain the structure and features of the chain $X^{(t)}$.

2.6.2.3 Limiting Distribution of Chain

Generally, it was shown that with the use of one-step transition probability matrix P and distribution for the initial status of the chain, the sequence of distributions $\left\{\Pi^{(t)}\right\}_{t \geq 0}$ is measurable for chain $\left\{X^{(t)}, t \geq 0\right\}$.

With all these explanations, if

$$\lim_{t \to \infty} \Pi^{(t)} = \Pi \quad (2.39)$$

where Π is a probability distribution, then Π is called the limiting distribution of chain $\left\{X^{(t)}, t \geq 0\right\}$. It is clear that for different choices of $\Pi^{(0)}$, the limiting distributions may differ, if any.

Definition 2.6.2

A Markov chain is said to be an ergodic chain if it is possible to move from every state to another state (not necessarily in one movement). In other words, the limiting distribution for any distributions sequence $\left\{\Pi^{(t)}\right\}_{t \geq 0}$ converges to distribution Π. That is,

$$\forall \Pi^{(0)} : \lim_{t \to \infty} \Pi^{(t)} = \Pi \tag{2.40}$$

In summary, regardless of the initial distribution of the chain, the limiting distribution in the ergodic chain is unique.

It is worth noting that this characteristic of the Markov chain will provide a method for sampling. Suppose we are going to extract a sample from the density function $f(x)$ with domain S, as well as suppose other typical methods applying for a sampling of this density function don't work well.

If we have an ergodic Markov chain, whose unique limiting distribution is $f(x)$, then if from each row of the one-step transition probability matrix (P) a sample can be generated, then we use the following algorithm to generate the total sample:

1. Generate $X^{(0)}$ from an arbitrary $\Pi^{(0)}$;
2. With having $X^{(t)} = x^{(t)}$ (for $t = 0, 1, 2, \ldots, n$), any sample unit of the next step ($X^{(t+1)} = x^{(t+1)}$) can be simulated by considering the previous row ($x^{(t)}$) of the probability transition matrix P;
3. Apply this algorithm to get the desired sample size.

In this case, due to the ergodicity of the chain, the distribution of $X^{(t)}$, for enough large t, leans toward $f(x)$. To improve the quality of the sample, as well as to reduce the impact of the choice of initial distribution, a very large sample in the above manner shall be generated and only the observed end-points shall be considered as the real sample. Additionally, to avoid creating a biased sample, the initial observations shall be discarded.

Definition 2.6.3

If we consider the distribution $\Lambda = [\lambda_i]_{i \in S}$ for the initial status of the Markov chain, in which

$$\Lambda^{(T)} = \Lambda^{(T)} \times P \tag{2.41}$$

then for $t \geq 0$, the distribution of $X^{(t)}$ is Λ, too. In this case, Λ is referred to as the stationary distribution for chain $\{X^{(t)}\}$ or, correspondingly, Λ is the stationary distribution for P.

Definition 2.6.4

For any $i, j \in S$, it can be said that state j is accessible from state i if

Quality and Reliability

$$\exists k \in N; \; p_{ij}^{(k)} = \Pr\left(X^{(t+k)} = j | X^{(t)} = i\right) > 0 \quad (2.42)$$

In other words, state is accessible from state if, after a finite time (steps), the state be achieved with the start from the state. Furthermore, the states and are called **communicate**, if and only if the state is accessible from state and state j is accessible from the state i.

Definition 2.6.5

$\{X^{(t)}\}$ is said to be an irreducible chain if all of its states are communicated.

Definition 2.6.6

For each $i \in S$, the period of the state i is indicated by notation $d(i)$ and is defined as follows:

$$d(i) = \gcd\{k \in N : p_{ii}^{(k)} > 0\} \quad (2.43)$$

in which gcd is the abbreviation of the greatest common divisor. Note that if $d(i) = 1$, then the state i is aperiodic. In other words, the state i is said to be aperiodic if i can return to i again after any steps and, generally, it is called the chain is aperiodic if any state is aperiodic.

Definition 2.6.7

The Markov chain $\{X^{(t)}\}$ is said to be a time-reversible Markov chain if there exists a distribution $\Pi = [\pi_i]_{i \in S}$ in which

$$\forall \; i, j \in S; \; \pi_i p_{ij} = \pi_j p_{ji} \quad (2.44)$$

If Markov chain $\{X^{(t)}\}$ is time-reversible, then $\Pi = [\pi_i]_{i \in S}$ is the stationary distribution and will be unique. This result will be derived from the summation of both sides of the above equation. That is,

$$\forall j \in S : \sum_{i \in S} \pi_i p_{ij} = \pi_j \quad (2.45)$$

And this means

$$\Pi^{(T)} = \Pi^{(T)} \times P \quad (2.46)$$

The result that comes up from these definitions is relevant to the ergodicity condition of a chain. In fact, if the Markov chain is irreducible, aperiodic, and time-reversible, then it is ergodic, too.

Notwithstanding several methods such as importance sampling, acceptance and rejection, the inverse of the distribution function, and the variable change methods, because of lack of applicability of the said methods especially in complex functions, sampling from density function is yet the fundamental problem in the Monte Carlo method. With having all explanations, we are now looking for powerful methods for extracting a sample with the lowest restrictions.

During the discussion of the ergodic chain, we introduced a method that facilitates the sample extraction of target function $f(x)$. In doing so, it was enough to have an ergodic Markov chain with a unique stationary limiting distribution $f(x)$ so that every row of its probability transition matrix can be sampled.

Indeed, the use of the MCMC method has provided such circumstances to obtain a large enough sample for approximating some numerical problems such as complex integrals. Hence, in doing this algorithm, finding an irreducible, aperiodic, and time-reversible Markov chain with unique and stationary density function is the basic issue.

To produce such a chain, two methods known as Metropolis–Hastings and Gibbs Sampler have been outlined.

2.6.3 Metropolis–Hastings Algorithm

One of the main problems in applying the Monte Carlo method to integral approximation is the inability of different methods for sampling from the target density function, and this matter becomes important when the function used is a complex. Thereby, the basis for the emergence of the MCMC methods was rooted in this issue. Because they face such complex functions during their research, this issue was more pursued by mathematicians and physicists.

Metropolis and Ulam (1949), Metropolis et al. (1395), Hastings (1970), and Chib and Greenberg (1995) presented the early solutions to this problem.

In Metropolis–Hastings algorithm with a selection of kernel of transitions probability $g(y \mid x)$ on domain S, and by having $X^{(t)}$ as an observation, this method will be able to generate new observation $X^{(t+1)}$ after three steps. It is worth mentioning that the existence of an initial value (also known as seed value) plays a principal role in proceeding with this algorithm.

At time $t = 0$, the $X^{(0)} = x^{(0)}$ will be extracted from an arbitrary distribution $\Pi^{(0)}$ with density function $\pi^{(0)}(x)$ which is defined on S, provided that $f\left(x^{(0)}\right) > 0$. Of course the amount of $x^{(0)} \in S$ can be selected arbitrarily provided that that $f\left(x^{(0)}\right) > 0$.

After having the initial value of this chain ($X^{(0)} = x^{(0)}$), the next new value of the chain can be gained by applying the following steps:

1. Generate the suggested amount X^* from $g(. \mid x^{(t)})$;
2. Calculate the ratio of the Metropolis–Hastings for $\left(x^{(t)}, X^*\right)$ which is defined as follows:

$$R(u, v) = \frac{f(v)g(u \mid v)}{f(u)g(v \mid u)} \quad (2.47)$$

3. Choose X^{t+1} with the use of following instruction:
   - $X^{(t+1)} = x^*$ with the probability $\min\{1, R(x^{(t)} = X^*)\}$;
   - $X^{(t+1)} = x^{(t)}$ with the probability $1 - \min\{1, R(x^{(t)} = X^*)\}$:

Continue this algorithm until when reaching the desired sample size.

If in the Metropolis–Hastings algorithm, the $g(y|x)$ get selected in such a way that

$$g(y|x) = g(y) \tag{2.48}$$

so that $g(.)$ is a density function on S, then any suggested x^* will depend on the previous sample unit, $x^{(t)}$, and will be generated from $g(y)$. Hence, the chain that is obtained by applying this algorithm will be independent.

With doing so, the ratio of Metropolis–Hastings can be rewritten as follows:

$$R(x^{(t)}, X^*) = \frac{f(X^*)g(x^{(t)})}{f(x^{(t)})g(X^*)} \tag{2.49}$$

Further, for any $x \in S$, in which $f(x) > 0$, if the amount of $g(x) > 0$, then the chain will be irreducible and periodic. Consequently, this chain will be ergodic with unique and stationary limiting distribution $f(x)$.

2.6.3.1 Gibbs Sampling Method

As mentioned earlier, for the approximation of the integral with the help of the Monte Carlo method, when it is not possible to extract a sample from the target density function $f(x)$, the Metropolis–Hastings algorithm is a general method for generating an ergodic chain with a unique density $f(x)$, in which the sample units of this chain can be simulated from $f(x)$.

But the implementation of this algorithm is possible when an appropriate transitions probability kernel (suggested density) is available to generate new observations provided that the previous observations are applied.

In the Metropolis–Hastings algorithm, another group of the chains that are able to create quality samples is known as independent chains. Now, we will outline another algorithm for solving the previous sampling problem.

Suppose $X = (X_1, X_2, X_3, \ldots, X_p) \sim f(x)$ and also the vector X_{-i} is defined as follows:

$$X_{-i} = (X_1, X_2, X_3, \ldots, X_{i-1}, X_{i+1}, \ldots, X_p); i = 1, 2, \ldots, p$$

As such, for $i = 1, 2, \ldots, p$ the conditional distribution of $X_i | X_{-i} = x_{-i}$ and its density function, i.e., $f(x_i|x_{-i})$, is available.

To apply the Gibbs sampling method, with an initial value of $x^{(0)}$, which is also referred to as "seed", the following algorithm needs to be run:

1. Choose a combination of components of $x_{(t)}$;
2. For $i = 1, 2, \ldots, p$, extract the values of X_i^* from the density function $f\left(x_i \big| X_{-i}^{(t)}\right)$ to complete the vector of $X^* = \left(X_1^*, X_2^*, X_3^*, \ldots, X_p^*\right)$;
3. Set $X^{(t+1)} = X^*$.

This algorithm has been developed by Gibbs to fortify the approximation of the calculation of complex integrals. Although this algorithm is rooted in the works of Geman and Geman (1984), because of the remarkable endeavors of Gibbs, this algorithm is now known as the Gibbs sampling method which is commonly used in the field of simulation.

2.6.3.2 Some Features of the Gibbs Sampling Method

The above algorithm would be efficient when its features cover all features of the Metropolis–Hasting algorithm. The properties of this algorithm are as follows:

A. The chain of $\left\{X^{(t)}\right\}_{t \geq 0}$ is Markov because $X^{(t+1)}$ only depends on $X^{(t)}$;
B. In the second stage, by the step-by-step generation of p components, i.e., X_i^*, of the vector X^*, the vector X^* will be simulated instead of direct generation of X^* from density $f\left(x_i \big| X_{-i}^{(t)}\right)$. To avoid the impact of the order of the generation of the components, at the first step one combination among $p!$ permutation of components of $X^{(t)}$ shall be selected and the second stage of the simulation should be followed in this selected order;
C. Performing the second step with respect to the order selected in step one is called "cycle";
D. For a specific cycle, with having $X^{(t)}$ in the ith step, the suggested value can generate $X^* = \left(x_1^{(t)}, x_2^{(t)}, \ldots, x_i^*, \ldots, x_p^{(t)}\right)$ from density $g_i(X^* \mid x^{(t)})$ in which

$$g_i\left(x^* \big| x^{(t)}\right) = \begin{cases} f\left(x_i^* \big| x_{-i}^{(t)}\right); & x_{-i}^* = x_{-i}^{(t)} \\ 0; & \text{otherwise} \end{cases} \quad (2.50)$$

In this case, the value of the Metropolis–Hastings ratio is equal to one as follows:

$$R\left(X^{(t)}, X^*\right) = \frac{f\left(x^*\right) g_i(x^{(t)} \mid x^*)}{f\left(x^{(t)}\right) g_i(x^* \mid x^{(t)})} = 1$$

Therefore, the suggested value X^* can be accepted in any cycle and also the Gibbs sampling method can be considered as a specific state of the Metropolis–Hastings algorithm.

E. Clearly, $f(x_i | x_{-i}) = c_i f(x)$ in which $c_i^{-1} = \int f(x_{-i}) \mathrm{d}x_{-i}$. Because each algorithm contains p sub-algorithms of Metropolis–Hastings so, even if in some cases the value of c_i is unknown, the Gibbs sampling algorithm will run without any problem.

F. To optimize the result of the algorithm, in a cycle where the first suggested component, X_1^*, is generated with respect to $X^{(t)}$, then the second suggested component X_2^* is better to be generated by using the density function $g_i(x^* \mid x_1^{(t+1)}, x_2^{(t)}, \ldots, x_p^{(t)})$ instead of applying the following density function:

$$g_i(x^* \mid x_1^{(t)}, x_2^{(t)}, \ldots, x_p^{(t)}) \qquad (2.51)$$

Likewise, for other components until the completion of the cycle, the new value $x_i^{(t+1)}$ shall be used instead of $x_i^{(t)}$.

G. The Markov chain produced by this algorithm is irreducible, time-reversible, aperiodic, and therefore ergodic in its wake. In fact, this algorithm is one of the applicable tools of the MCMC method.

2.7 SLICE SAMPLING

Slice sampling is one of the specific states of the MCMC simulation and is based on uniform sampling from the surface beneath the conditional density curve or a pro-rate function thereto. As a result, applying this method only requires a simulation of the uniform distribution. This method is often used when the domain of the density function is finite or can be approximated in a finite area. The simplest model of slice sampling method is as follows.

Suppose $f(x)$ is the complete conditional density function that is used to generate a sample. Additionally, consider the density function $g(x)$ as proportional function to the target function $f(x)$, in which $g(x) = cf(x)$. To apply this method, we must uniformly sample from the two-dimensional area under curve $g(x)$. Therefore, for this purpose, it is necessary to use an auxiliary variable (like Y) with the following conditional distribution:

$$Y \mid X \sim \text{Uniform}(0, g(x)) \qquad (2.52)$$

In these circumstances, the variable (X, Y) has a joint density function on the area of $\{(x, y); 0 < y < g(x)\}$ as follows:

$$f(x, y) = f(y \mid x) f(x) = \frac{1}{c} I_{\{0 < y < g(x)\}}(x, y) \qquad (2.53)$$

On the other hand,

$$f(x \mid y) \propto f(x, y) = \frac{1}{c} I_{\{0 < y < g(x)\}}(x, y) \qquad (2.54)$$

This means $X \mid Y$ has a uniform distribution on the area of $S(y)$ as follows:

$$X \mid Y \sim \text{Uniform}(0, S(y)), \quad S(y) = \{x : g(x) \geq y\}$$

In fact, $X \mid Y$ has a uniform distribution on a slice of function $g(x)$ that is interrupted with line $Y = y$. For this reason, this method of sampling is called the "slice sampling method".

Since the conditional distributions are of known uniform distributions with known parameters, with initial values $\left(x^{(0)}, y^{(0)}\right)$ and by applying the Gibbs sampling algorithm, the subsequent new values of $\left(x^{(t)}, y^{(t)}\right)$ can be easily calculable as follows:

1. Generate the value of $y^{(t)}$ from the below density function,

$$f\left(y|x^{(t-1)}\right) = \text{Uniform}\left(0, \, g\left(x^{(t-1)}\right)\right) \tag{2.55}$$

2. Generate the value of $x^{(t)}$ from the below density function,

$$f\left(x|y^{(t)}\right) = \text{Uniform}\left(0, \, S\left(y^{(t)}\right)\right) \tag{2.56}$$

3. After the convergence of the chain $\{x^{(t)}, y^{(t)}\}$, the values of $y^{(t)}$ should be eliminated to reach a simulated sample of function $f(x)$. The simulation procedure of the slice sampling method is similar to the data-generation way using the CDF method.

To use the programming codes of the slice sampling method, see Section B.3 in the Appendix.

Given that the Gibbs sampling method will also be used here, it was necessary to briefly discuss the Markov Monte Carlo method.

Since we need to use several concepts and terms in this book, the features of reliability data, different types of censorships, and ALT methods will be discussed in the coming chapters. Given that the Gibbs sampling method will also be used here, it was necessary to briefly discuss the Markov Monte Carlo method.

In this chapter, we first get acquainted with the Dirichlet distribution. Then, we review the Dirichlet process and some of its features. After understanding this process and how it changes with respect to different parameters, in Section 2.2, we examine Pólya's Urn Model and its relation to the Dirichlet process. In Section 2.3, we introduce a prediction method using the Blackwell–Macqueen Urn Scheme, and finally, we examine one of the features and capabilities of the Dirichlet process, namely data clustering using the Chinese Restaurant Process (CRP).

3 Dirichlet Process

As mentioned, Ferguson (1973) introduced the Dirichlet process as a prior distribution for all distribution functions and discussed two important features of this process (namely the support of all distribution functions and its conjugation).

This process is a noninformative prior distribution, and this feature makes it easy to calculate with and also, due to the least constraint on the parameters, lead to a more accurate inference.

The development of Bayesian computational methods, such as the MCMC method, has led to the use of the Dirichlet process as a prior distribution in nonparametric Bayesian problems, nonparametric regression, and nonparametric density estimation.

As regards the issues mentioned in the preface section regarding the Dirichlet process and its background, in order to avoid repetitive content, this chapter provides a thorough review of the background and application of this process.

3.1 DIRICHLET DISTRIBUTION

For a better understanding of the Dirichlet distribution, we present the following example.

Suppose random variable X takes values 1 and 2 with probabilities p and $1-p$, respectively. As you know, in Bayesian inference, estimation of the unknown parameter p will be computed by applying an appropriate prior density representing the features of this parameter. The use of the Bayesian method will finally lead to the creation of the posterior function that is applicable in the estimation of unknown parameters (like p herein).

A suitable prior density function for parameter p in this example is as follows:

$$\pi(p) \propto p^{a_1-1}(1-p)^{a_2-1} \tag{3.1}$$

That is a beta distribution with parameters (a_1, a_2).

With the choice of values $a_i = \alpha m_i$ for a_1 and a_2 in which α is a positive and real value so that

$$0 < m_i < 1$$

and

$$m_1 + m_2 = 1$$

The following results can be gained from this example:

$$E(P) = \frac{\alpha m_1}{\alpha m_1 + \alpha m_2} = m_1$$

$$\mathrm{Var}(P) = \frac{\alpha^2 m_1 m_2}{(\alpha m_1 + \alpha m_2)^2 \times (\alpha m_1 + \alpha m_2 + 1)} = \frac{m_1(1-m_1)}{\alpha+1} \quad (3.2)$$

If we set $m_1 = 0.5$ and $\alpha = 2$, the prior distribution in equation (3.2) will convert to a noninformative prior, namely Uniform $(0,1)$. The value m_1 is the expected value of the variable P, and α affects the precision of the prior distribution. That means as much the α gets bigger the variance decreases and the parameter p will be dispersed and concentrated near the expected/mean value m_1. Take heed another example.

Consider a Bernoulli experiment with N repeat in which $X = 1$ and $X = 2$ are repeated n_1 and n_1 times $(n_1 + n_2 = N)$, respectively. With the assumption that $X_1, X_2, \ldots, X_N$ are the variables representing the result of the relevant experiment, the posterior distribution of parameter p, with the Beta(a_1, a_2) prior distribution, is as follows:

$$\pi(p \mid X_1, X_2, \ldots, X_N) \propto p^{n_1}(1-p)^{n_2} p^{a_1-1}(1-p)^{a_2-1}$$
$$\propto p^{n_1+a_1-1}(1-p)^{n_2+a_2-1} \quad (3.3)$$

This is a beta distribution with parameters $n_1 + a_1$ and $n_2 + a_2$. Clearly, since the posterior distribution belongs to the same family of the prior distribution (namely the beta distribution), the distribution in equation (3.2) is a conjugate prior. The mentioned example is known as the beta-binomial model for parameter p. With the generalization of the beta-binomial model, the Dirichlet-Binomial model is achieved.

To get more understanding about this model, suppose variable X takes values x_i for $i = 1, 2, \ldots, n$ with probabilities p_i in which $0 \le p_i \le 1$ and $\sum_{i=1}^{n} p_i = 1$. By generalizing the model in (2.1), a suitable prior distribution for the vector of unknown parameters $(p_1, p_2, \ldots, p_n)$ is as follows:

$$\pi((p_1, p_2, \ldots, p_n) \propto p_1^{a_1-1} \times p_2^{a_2-1} \times \ldots \times p_n^{a_n-1} \quad (3.4)$$

This density function demonstrates a Dirichlet distribution. In general, the vector $(p_1, p_2, \ldots, p_n)$ has the Dirichlet distribution if the form of its probability density is as follows:

$$\pi(p_1, p_2, \ldots, p_n) = \frac{\Gamma(a_1 + a_2 + \ldots + a_n)}{\prod_{i=1}^{n}(a_i)} \prod_{i=1}^{n} p_i^{a_i-1} \quad (3.5)$$

where

$$0 \le p_i \le 1 \text{ and } \sum_{i=1}^{n} p_i = 1$$

And we write $(p_1, p_2, \ldots, p_n) \sim \mathrm{Dir}(a_1, a_2, \ldots, a_n)$, in which $\Gamma(.)$ denotes the gamma function. As a matter of fact, it can be said that the Dirichlet distribution is a distribution over distributions. That means, the Dirichlet distribution covers other distributions.

As said previously, if we set $a_i = \alpha m_i$ so that $0 < m_i < 1$ and $\sum_{i=1}^{n} m_i = 1$, then the Dirichlet distribution can be written as follows:

$$\pi(p_1, p_2, \ldots, p_n) = \frac{\Gamma(\alpha)}{\prod_{i=1}^{n} (\alpha m_i)} \prod_{i=1}^{n} p_i^{\alpha m_i - 1} \tag{3.6}$$

Similar to the foregoing example, the expected value for P_i and the covariance of P_i and P_j can be calculated as follows:

$$E(P_i) = \frac{\alpha m_i}{\alpha \sum_{i=1}^{n} m_i} = m_i$$

$$\mathrm{Var}(P_i) = \frac{\alpha m_i (\alpha - \alpha m_i)}{\alpha^2 (\alpha + 1)} = \frac{m_i (1 - m_i)}{(\alpha + 1)} \tag{3.7}$$

$$\mathrm{Cov}(P_i, P_j) = \frac{-\alpha m_i \alpha m_j}{\alpha^2 (\alpha + 1)} = \frac{-m_i m_j}{(\alpha + 1)}, \ i \neq j$$

in which the unknown parameter α is known as precision or concentration parameter that controls the dispersal of the Dirichlet distribution around the expected value. Therefore, to sum it all up, the Dirichlet distribution can be considered as a conjugate prior to the binomial distribution.

To more explanation of the Dirichlet distribution, suppose $\chi = \{\chi_1, \chi_2, \ldots, \chi_n\}$ shows n discrete space with corresponding probabilities in the set of $\Theta = \{\theta_1, \theta_2, \ldots, \theta_n\}$ so that $\Pr\{X = \chi_i\} = \theta_i$.

In such circumstances, it can be said that the prior distribution for the vector of parameters is Dirichlet. That means $\Theta \sim \mathrm{Dir}(\Theta, \alpha, M)$ and its probability density function can be written as follows:

$$\pi(\theta_1, \theta_2, \ldots, \theta_n) = \frac{\Gamma(\alpha)}{\prod_{i=1}^{n} (\alpha m_i)} \prod_{i=1}^{n} \theta_i^{\alpha m_i - 1}, \ a > 0 \tag{3.8}$$

where $M = \{m_1, m_2, \ldots, m_n\}$ is known as the base measure of the probability space χ. It is worth mentioning that if we set $\alpha = 1$, then the density function in (2.4) will convert to the uniform distribution.

Another way to build the Dirichlet distribution is by applying the gamma distribution. Suppose the random variable Z comes from a gamma distribution with parameters (α, β) if its probability density function is as follows:

$$f(z|\alpha, \beta) = \frac{1}{\Gamma(\alpha) \beta^\alpha} z^{\alpha - 1} e^{-\frac{z}{\beta}} I_{(0,\infty)}(z) \tag{3.9}$$

where $I_s(z)$ is a notation of indicator function that is defined on the domain s. Namely, $I_s(z) = 1$ if $z \in s$ and is zero otherwise.

To make the Dirichlet distribution by applying the gamma distribution, suppose $Z_1, Z_2, \ldots, Z_n$ are independent variables with distinct distributions of Gamma $(\alpha_i, 1)$. Clearly,

$$\sum_{i=1}^{n} Z_i \sim \text{Gamma}\left(\sum_{i=1}^{n} \alpha_i, 1\right), \quad \alpha_i > 0 \qquad (3.10)$$

Now, according to the variable defined by Ferguson (1973) which is

$$Y_i = \frac{Z_i}{\sum_{i=1}^{n} Z_j} \qquad (3.11)$$

it is easy to find that $(Y_1, Y_2, \ldots, Y_n) \sim \text{Dir}(\alpha_1, \alpha_2, \ldots, \alpha_n)$. For example, if we consider $n = 2$, then the Dirichlet distribution will convert to the beta distribution with parameters α_1 and α_2.

To get a better understanding of this distribution and its properties, pay attention to the following examples.

Example 3.1.1

To investigate the behavior of the three-dimensional Dirichlet distribution, we simulate a sample of a volume of 1000 units of this distribution with different precision parameters. The scatter plot of this sample is shown in Figure 3.1. The top-left graph relates to $\text{Dir}(0.1, 0.1, 0.1)$, which shows these 1000 sample units are dispersed in the margin of the distribution and are away from the expected value. The top-right graph relates to $\text{Dir}(1,1,1)$. As this figure shows, the behavior of these data is similar to the uniform distribution throughout the space. The bottom-left chart relates to $\text{Dir}(10,10,10)$. Since the precision parameter of these data is more than one, as the graph shows, the data is expected to concentrate close to the mean value. Finally, the bottom-right figure relates to $\text{Dir}(0.1,1,10)$, which shows the skewness in the Dirichlet distribution.

This graph shows that the Dirichlet distribution is symmetric and skewed like the beta distribution and, as a result, indicates that the Dirichlet distribution is highly flexible. To use the programming codes of this graph, see Section B.4 in the Appendix.

Example 3.1.2

For better intuition, in Figure 3.2, we have drawn another diagram of the Dirichlet distribution behavior for different parameter values. In this chart, the yellow color indicates the nearness of the data to the mean value and the red color refers to the data on the margin and away from the mean.

If we arrange the graphs from the top-left in the clockwise direction, the first figure pertains to $\text{Dir}(0.1, 0.1, 0.1)$. Due to the low amount of precision parameter

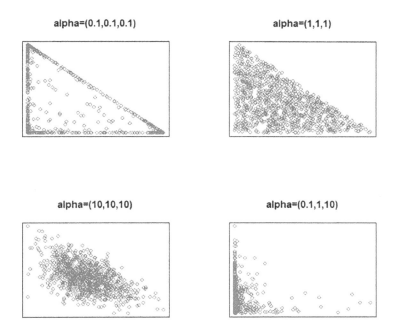

FIGURE 3.1 Simulated Dirichlet distribution with different precision parameters. (Authors' own figure.)

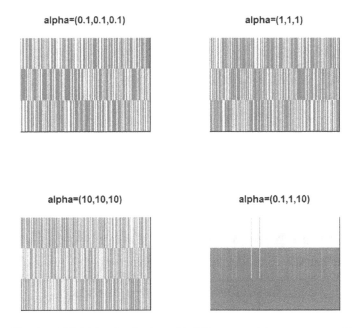

FIGURE 3.2 Simulated Dirichlet distribution with different precision parameters. (Authors' own figure.)

in this dataset, the data of this distribution have distant from the mean value and are in the margin of the space. The second graph relates to Dir(1,1,1). As mentioned in the previous example, the behavior of this specific distribution is similar to the uniform distribution. Likewise, the third graph reflects the dispersal of data gathered from Dir(10,10,10) around the mean value. As you observe, because of the high amount of precision parameters in this specific state of the Dirichlet distribution, the data concentrates around the mean value. Finally, the last graph pertains to the Dir(0.1,1,10), which refers to the remarkable flexibility of this distribution.

To use the programming codes of this graph, see Section B.5 in the Appendix.

3.1.1 Remarkable Properties of the Dirichlet Distribution

Since the Dirichlet distribution can also be obtained from the gamma distribution, some properties of these two distributions (especially in the collectivity property) are similar. Some other features of the Dirichlet distribution include:

A. Suppose $(Y_1, Y_2, \ldots, Y_n) \sim \mathrm{Dir}(\alpha_1, \alpha_2, \ldots, \alpha_n)$ and r_i for $i = 1, 2, \ldots, l$ are integers. So that $0 < r_1 < r_2 < \ldots < r_l = n$. Then:

$$\left(\sum_{i=1}^{r_1} Y_i, \sum_{i=r_1+1}^{r_2} Y_i, \ldots, \sum_{i=r_{l-1}+1}^{n} Y_i\right) \sim \mathrm{Dir}\left(\sum_{i=1}^{r_1} a_i, \sum_{i=r_1+1}^{r_2} a_i, \ldots, \sum_{i=r_{l-1}+1}^{n} a_i\right) \quad (3.12)$$

B. The marginal distribution of Y_i for $i = 1, 2, \ldots, n$ is beta. That is,

$$Y_i \sim \mathrm{Beta}\left(a_i, \sum_{j=1}^{n} a_j - a_i\right) \quad (3.13)$$

C. If $(Y_1, Y_2, \ldots, Y_n) \sim \mathrm{Dir}(\alpha_1, \alpha_2, \ldots, \alpha_n)$, then

$$E(Y_i) = \frac{a_i}{\alpha}$$

$$E(Y_i^2) = \frac{a_i(a_i+1)}{\alpha(\alpha+1)} \quad (3.14)$$

$$E(Y_i Y_j) = \frac{a_i a_j}{\alpha(\alpha+1)}, \quad i \neq j$$

in which $\alpha = \sum_{i=1}^{n} a_i$. Note that all features of this distribution referred to in A, B, and C are provable by using the properties of the gamma distribution.

The next features of this distribution, which have been listed in the following items, are applicable in the inference statistics.

D. If $(Y_1, Y_2, \ldots, Y_n) \sim \mathrm{Dir}(\alpha_1, \alpha_2, \ldots, \alpha_n)$ and for $j = 1, 2, \ldots, n$;

$$\Pr\{X = j | Y_1, Y_2, \ldots, Y_n\} = Y_j \tag{3.15}$$

Therefore, the posterior distribution of $Y_1, Y_2, \ldots, Y_n | X = j$ is $\text{Dir}\left(\alpha_1^{(j)}, \alpha_2^{(j)}, \ldots, \alpha_n^{(j)}\right)$ in which

$$\alpha_i^{(j)} = \begin{cases} \alpha_i & \text{if } i \neq j \\ \alpha_i + 1 & \text{if } i = j \end{cases} \tag{3.16}$$

To prove this, see Section A.1 in the Annex.

From this property of the Dirichlet distribution, a formula is obtained that will be used in future relationships. By considering the following equality,

$$\Pr\{X = j, Y_1 \leq z_1, \ldots, Y_n \leq z_n\} = \Pr\{X = j\}$$
$$\Pr\{Y_1 \leq z_1, \ldots, Y_n \leq z_n | X = j\} \tag{3.17}$$

If $\text{Dir}(y_1, y_2, \ldots, y_n | \alpha_1, \alpha_2, \ldots, \alpha_n)$ be a notation of Dirichlet distribution, the equation in (2.5) can be rewritten as follows:

$$\int_0^{z_1} \ldots \int_0^{z_n} y_j \text{dDir}(y_1, y_2, \ldots, y_n | \alpha_1, \alpha_2, \ldots, \alpha_n)$$
$$= \frac{a_j}{\alpha} \text{Dir}\left(z_1, z_2, \ldots, z_n | \alpha_1^{(j)}, \alpha_2^{(j)}, \ldots, \alpha_n^{(j)}\right) \tag{3.18}$$

Note that, this rewritten equation is true even if $a_j = 0$.

The left side of equation (3.17) can be rewritten as follows:

$$\Pr\{X = j, Y_1 \leq z_1, \ldots, Y_n \leq z_n\}$$
$$= \int_0^{z_1} \ldots \int_0^{z_n} \Pr\{X = j, Y_1 = y_1, \ldots, Y_n = y_n\} dy_1 \ldots dy_1$$
$$= \int_0^{z_1} \ldots \int_0^{z_n} \Pr\{X = j | Y_1 = y_1, \ldots, Y_n = y_n\} \text{dDir} \tag{3.19}$$
$$(y_1, y_2, \ldots, y_n | \alpha_1, \alpha_2, \ldots, \alpha_n)$$
$$= \int_0^{z_1} \ldots \int_0^{z_n} y_j d\text{Dir}(y_1, y_2, \ldots, y_n | \alpha_1, \alpha_2, \ldots, \alpha_n)$$

Likewise, on the right side of equation (3.17), $\Pr\{X = j\}$ can be rewritten as follows:

$$\Pr\{X = j\} = \int_0^1 \ldots \int_0^1 \Pr\{X = j, Y_1 = y_1, \ldots, Y_n = y_n\} dy_1 \ldots dy_1$$

$$= \int_0^1 \ldots \int_0^1 \Pr\{X = j \mid Y_1 = y_1, \ldots, Y_n = y_n\} \mathrm{dDir}$$

$$(y_1, y_2, \ldots, y_n \mid \alpha_1, \alpha_2, \ldots, \alpha_n) \qquad (3.20)$$

$$= \int_0^1 \ldots \int_0^1 y_j \mathrm{dDir}(y_1, y_2, \ldots, y_n \mid \alpha_1, \alpha_2, \ldots, \alpha_n)$$

$$= E(Y_j) = \frac{a_j}{\alpha}$$

E. Another feature of the Dirichlet distribution, as noted earlier, is that it can be considered to be a conjugate prior when the binomial distribution is applied as the present evidence.

In other words, suppose $X_1, X_2, \ldots, X_N$ is a vector of random variables with a discrete distribution that take values $x_1, x_2, \ldots, x_n$ with probabilities $p_1, p_2, \ldots, p_n$. Additionally, suppose the value x_i is observed n_i times where $\sum_{i=1}^{n} n_i = N$.

By considering the Dirichlet distribution as the prior distribution for the vector of parameters $p = (p_1, p_2, \ldots, p_n)$, the posterior distribution will be Dirichlet, too, as follows:

$$\pi(p \mid \alpha, X_1, X_2, \ldots, X_N) \propto f(X_1, X_2, \ldots, X_N \mid \alpha, p) \pi(p \mid \alpha)$$

$$\propto \prod_{i=1}^{n} p_i^{n_i} \prod_{i=1}^{n} p_i^{a_i - 1}$$

$$\propto \prod_{i=1}^{n} p_i^{n_i + a_i - 1} \qquad (3.21)$$

$$\propto \mathrm{Dir}(\alpha_1 + n_1, \alpha_2 + n_2, \ldots, \alpha_n + n_n)$$

The expected value of the posterior variable $p_i \mid X_1, X_2, \ldots, X_N$, for $i = 1, 2, \ldots, n$ can be calculated as follows:

Dirichlet Process

$$E(P_i | X_1, X_2, \ldots, X_N) = \frac{\alpha_i + n_i}{\sum_{i=1}^{n}(\alpha_i + n_i)}$$

$$= \frac{\alpha m_i + n_i}{\alpha \sum_{i=1}^{n} m_i + \sum_{i=1}^{n} n_i} \qquad (3.22)$$

$$= \frac{\alpha m_i + n_i}{\alpha + N}$$

$$= \frac{n_i}{\alpha + N} + \frac{\alpha m_i}{\alpha + N}.$$

In fact, the mean value of the posterior variable is a composition of the mean value of the prior distribution m_i and the number of repetitions n_i. This equality embraces interesting points. For instance:

1. When the sample size N increases, the fraction $\frac{\alpha m_i}{\alpha + N}$ leans toward zero. This signifies that if the sample size goes toward infinity, the information of present data overcomes the prior information;
2. If $\alpha \to \infty$ and N is constant, the prior information overcomes the information of present data;
3. If $\alpha \to 0$, the information of present data overcomes the prior information.

One particular model of the Dirichlet distribution is when it is two-dimensional (namely $n = 2$). Consider the following example for clarification.

Example 3.1.3

In the two-dimensional mode, the Dirichlet distribution will convert to the beta distribution with parameters (α, α). Figure 3.3 relates to the two-dimensional Dirichlet distribution. By considering $\alpha = 1$ in this distribution, the uniform distribution with known parameters (0,1) (i.e., Uniform(0,1)) will appear. Therefore, simulated data of Dir (1,1) is expected to treat like Uniform(0,1).

To use the programming codes of this graph, see Section B.6 in the Appendix.

With considering the explained assumptions above, suppose X_{N+1} is a variable with Dirichlet distribution that pertains to the $(N+1)$th observation. If χ_j is considered as an event space in the sample space of χ, the probability of $(N+1)$th observation given the previous observations and also the parameters of the Dirichlet distribution, α and M, can be calculated by applying marginal distribution as follows:

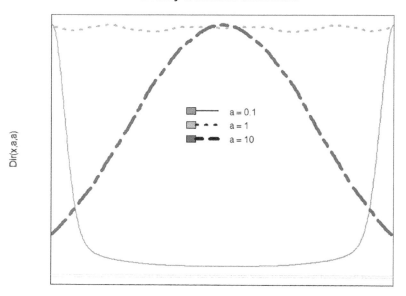

FIGURE 3.3 Probability density plot of two-dimensional Dirichlet distribution with a different precision parameter. (Authors' own figure.)

$$\Pr\{X_{N+1} \in \chi_j | X_{1:N}, \alpha, M\} = \int_\Theta \Pr\{X_{N+1} \in \chi_j | \Theta\} \Pr\{\Theta | X_{1:N}, \alpha, M\} d\Theta$$

$$= \int_\Theta \theta_j \mathrm{Dir}\left(\Theta | \alpha^*, M^*\right) d\Theta$$

$$= E(\theta_j) \quad (3.23)$$

$$= \frac{\alpha m_j^*}{\sum_{i=1}^n m_i^*}$$

$$= m_i^*$$

in which $\alpha^* = \alpha + N$ and $M^* = \{m_1^*, m_2^*, \ldots, m_n^*\}$. So that

$$m_j^* = \frac{\alpha m_j}{\alpha + N} + \frac{1}{\alpha + N} \sum_{i=1}^n \delta(x_i) \quad (3.24)$$

In summary,

$$\Pr\{X_{N+1} \in \chi_j | X_{1:N}, \alpha, M\} = \frac{\alpha m_j}{\alpha + N} + \frac{1}{\alpha + N} \sum_{i=1}^{n} \delta(x_i) \qquad (3.25)$$

Note that equation (3.25) results from one of the properties of the Dirichlet distribution namely, $E(\theta_j) = \frac{M(\chi_j)}{M(\chi)}$.

To prove this, see Section A.2 in the Annex.

This predictive probability is one of the important features of the Dirichlet distribution that plays a key role in the simulation of the posterior distribution.

3.2 DIRICHLET PROCESS

If we want to describe the Dirichlet process in a sentence, it can be said that it is a continuous state of Dirichlet distribution. That means, in the Dirichlet distribution, the given variable holds a distribution with discrete probability, whereas in the Dirichlet process, this distribution is a continuous one. What Ferguson (1973) did by using this process was taking this process as a prior distribution for the given infinite parameters.

Because the way used by Ferguson had a significant effect on the nonparametric Bayesian approach, this process was identified as the base of the nonparametric Bayesian models.

a. Ferguson (1973) considered two important points for the prior distribution of parameters. Firstly, since in the nonparametric Bayesian the number of parameters is infinite, the selected prior distribution should cover all the distributions. That is, the selected prior distribution of parameters should not be specific to a particular family of distributions because otherwise, and some restrictions have been considered on parameter estimation.
b. Secondly, since the number of parameters is high, calculating the posterior distribution of the parameters using Bayes' law will be complicated, and this will cause the posterior distribution of parameters not to be a closed form. So the selected prior distribution, in addition to covering all distributions, should have a simple form rather than a complex structure.

The Dirichlet process not only is a conjugate distribution to the binomial distribution family but also, with probability one, covers all discrete distributions. In the previous section, we found that when the number of parameters in the binomial family is finite, the Dirichlet distribution can be considered as a conjugate prior distribution to the existing parameters.

Now, if the number of the binomial distribution parameters is not finite, can the Dirichlet distribution still play the prior role for the parameters? Our goal in this section is to respond to such questions and introduce some of the features of the Dirichlet process.

Since the Dirichlet process is a probability measure over the space of all discrete distribution functions, it is necessary to have information on the probability domain in infinite-dimension spaces, cylinders, Borel cylinders, and the Kolmogorov extension theorem. These are mentioned in Chapter 3 of the book written by Kolmogorov (1933). In order to avoid duplication of content, we will state some of the theorems used here and refer the reader to the relevant sources for proof.

Suppose χ is a sample space and β refers to the Borel sigma-algebra of space χ. According to Ferguson (1973), consider the following definition.

Definition 3.2.1

Suppose α is a non-null infinite measure on space of (χ, β). So G has a Dirichlet process on (χ, β) with precision parameter α if; for each partition of χ like $(B_1, B_2, ..., B_k)$, for $i = 1, 2, 3, ...$, the vector $(G(B_1), G(B_2), ..., G(B_k))$ has a Dirichlet distribution. That means if

$$\left(G(B_1), G(B_2), ..., G(B_k)\right) \sim \mathrm{Dir}\left(\alpha(B_1), \alpha(B_2), ..., \alpha(B_k)\right) \qquad (3.26)$$

Since α is a non-null infinite measure on space of (χ, β), in some texts, it is deemed to be equated with αG_0 where G_0 is a probability measure on (χ, β) and α is a positive real number. In fact, if G has a Dirichlet process, it is symbolized with $G \sim \mathrm{DP}(\alpha)$, and if αG_0 replaced α, it is notated with $G \sim \mathrm{DP}(G_0, \alpha)$.

Theorem 3.2.1

Suppose G has a Dirichlet process on (χ, β) with parameters α and G_0, and also consider $A \in \beta$. If $G_0(A) = 0$, then with probability one $G(A) = 0$, and if $G_0 > 0$, then with probability one $G(A) > 0$. Additionally,

$$E\big(G(A)\big) = G_0(A)$$
$$\mathrm{Var}\big(G(A)\big) = \frac{G_0(A)\big(1 - G_0(A)\big)}{1 + \alpha} \qquad (3.27)$$

To prove this, see Section A.3.

From the expected value and the variance of this process, it is easy to see that this process is the generalized version of the Dirichlet distribution. By considering the Dirichlet distribution as a specific state of the n-dimensional process, Ferguson (1973) was able to acquire the Dirichlet process by the generalization of the Dirichlet distribution.

As it is shown, the Dirichlet process like the Dirichlet process contains two parameters including the precision or concentration parameter α and the base distribution G_0. Like in the Dirichlet distribution, these parameters play a key role in the Dirichlet process, too. That means as the parameter α increases, the distribution of G approaches the base distribution G_0 (and vice versa). It is worth mentioning that the Dirichlet process is often referred to as "the distribution over distributions".

Dirichlet Process

According to the definition of the Dirichlet process by Ferguson (1973), it can be said that for any $B \in \beta$,

$$G(B) \sim \beta\left(\alpha G_0(B),\ \alpha(1-G_0(B))\right) \qquad (3.28)$$

For more details, see Examples 2.3.1 and 2.3.2.

Example 3.2.1

Suppose $G \sim DP(G_0, \alpha)$ in which the base distribution G_0 is a Weibull distribution with known parameters $(\lambda, \beta) = (2, 3)$. We simulate a set of data of this process for different amounts of the precision parameter α. Figure 3.4 illustrates the drawn plot of this process with the Weibull base distribution (which is shown with "+" signs in blue). The Weibull cumulative distribution with mentioned known parameters (which is shown with black circle signs) has been drawn, too. The sample size of both charts is 400.

The interpretation of the charts from top-left in clockwise is as follows.

The top-left chart is related to this process with $\alpha = 0.1$. As you see, like the Dirichlet distribution, when α is greater than one, the data is dispersed on marginal and does not tend to be close to the mean value. The second figure relates to simulated data from the same process whose concentration parameter is equal to one. The diagram shows that the Dirichlet process behaves like a Dirichlet distribution

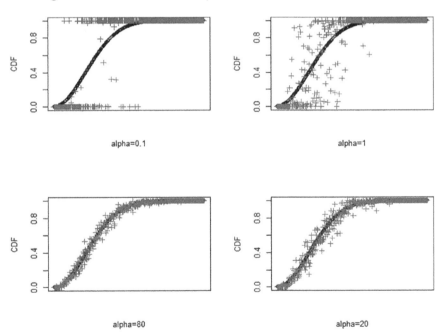

FIGURE 3.4 The cumulative distribution function of simulated data generated from the Dirichlet process with the Weibull base distribution with known parameters for a different precision parameter. (Authors' own figure.)

when the concentration parameter is equal to one and may some data disperse around the base distribution (the base distribution G_0 is Weibull).

However, the last two diagrams illustrate the simulated data from the Dirichlet process with the concentration parameters 20 and 80, respectively. It is clear that these data have a severe tendency to fit on the Weibull distribution. That is, the data is strongly centered around the mean so that the last graph almost refers to this fact that the simulated data from the Dirichlet distribution, with base distribution Weibull, is completely fitted on the Weibull distribution with known parameters $(\lambda, \beta) = (2, 3)$.

To use the programming codes of this figure, see Section B.7 in the Appendix.

For further understanding of how data from the Dirichlet process are affected by the decision parameter as well as the base distribution, consider the following example.

Example 3.2.2

Suppose a Dirichlet process with a known base distribution and different values of the precision parameters. The Lomax distribution with known parameters $(\alpha, \beta) = (3.2, 2.5)$ is considered as the base distribution. Figure 3.5 results from 400 data generated from this process with the known base distribution for different

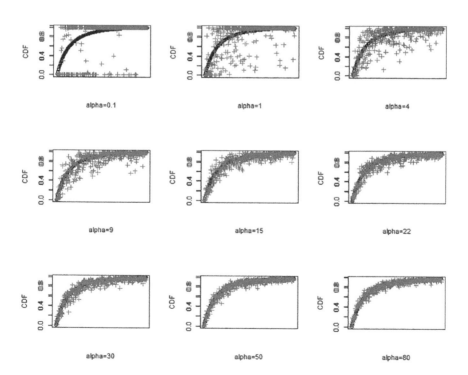

FIGURE 3.5 The cumulative distribution function of simulated data generated from the Dirichlet process with the Lomax base distribution with known parameters for a different precision parameter. (Authors' own figure.)

Dirichlet Process

precision parameters. In summary, the interpretation of these nine graphs in order of clockwise with the start of the top-left graph is as follows.

As this graph shows, with the growth of the precision parameter from 0.1 to 80, the tendency of the generated data from this process toward the Lomax distribution, as the base distribution, is apparent. So that the data of this process with $\alpha = 80$ almost corresponds to the data generated from the Lomax distribution with known parameters $(\alpha, \beta) = (3.2, 2.5)$.

By drawing Figures 3.7 and 3.8, it is expected that the relationship between the precision parameter and the base distribution in the Dirichlet process is understandable. To use the programming codes of Figure 3.8, see Section B.8 in the Appendix.

According to Görür and Rasmussen (2010), Properties of the Dirichlet process and its sensitivity to the concentration/precision and distribution parameters have made it important in Bayesian analyzes. The importance of this fact is undeniable especially when the selection of the prior distribution for base distribution is under question.

Definition 3.2.2

Suppose G is a random probability measure on (χ, β). $X_1, X_2, \ldots, X_n$ is a sample of random variables of G if for $m = 1, 2, \ldots$, and measurable sets $A_1, A_2, \ldots, A_m$ and $C_1, C_2, \ldots, C_n$, the following relationship comes true with probability one:

$$\Pr\{X_1 \in C_1, \ldots, X_n \in C_n | G(A_1), \ldots, G(A_m), G(C_1), \ldots, G(C_n)\}$$

$$= \prod_{j=1}^{n} G(C_j) \qquad (3.29)$$

In other words, $X_1, X_2, \ldots, X_n$ is a sample of random variables of G if the sets $\{X_1 \in C_1\}, \ldots, \{X_n \in C_n\}$ are independent of the rest of the process. Additionally, these sets are independent of each other, so that

$$\Pr\{X_j \in C_j | G(C_1), \ldots, G(C_n)\} = G(C_j) \qquad j = 1, 2, \ldots, n \qquad (3.30)$$

The above relationship is summarized as follows:

$$X_1, X_2, \ldots, X_n | G \sim G \qquad (3.31)$$

This model and symbol will be used widely in the next chapter to model the hierarchical form of the data simulation algorithm.

Theorem 3.2.2

Suppose G is a Dirichlet process on the measurable space (χ, β) with the given α parameter. Consider the X as a variable with distribution G. So, for any subset $A \in \beta$:

$$\Pr\{X \in A\} = \frac{\alpha(A)}{\alpha(\chi)} \qquad (3.32)$$

To prove this relation, see Section A.4 in the Appendix.

One important theorem that facilitates the simulation process in the nonparametric Bayesian inference using the Dirichlet process is the following theorem. This theorem clarifies the relationship between the posterior distribution, the prior distribution, and the empirical distribution as well as illustrates how the precision parameter affects the sampling process of the posterior distribution.

Theorem 3.2.3

Suppose G is a Dirichlet process on the measurable space (χ, β) with the precision parameter α and consider $X_1, X_2, \ldots, X_n$ is a sample of random variables of G. So, $G \mid X_1, X_2, \ldots, X_n$ is a Dirichlet process with precision parameter $\alpha + \sum_{i=1}^{n} \delta_{X_i}$. Additionally, for each measurable partition of χ, like $B_1, B_2, \ldots, B_k$:

$$(G(B_1), \ldots, G(B_k)) \mid X_1, X_2, \ldots, X_n \sim \text{Dir}(\alpha G_0(B_1) + n_1, \ldots, \alpha G_0(B_k) + n_k) \qquad (3.33)$$

where n_i is the number of observations in the set B_i.

If in Theorem 3.2.3, we use αG_0 in place of α, then the distribution of $G \mid X_1, X_2, \ldots, X_n$ is a Dirichlet process with parameter $\alpha^* G_0^*$ in which $\alpha^* = \alpha + n$ and

$$G_0^* = \frac{\alpha}{\alpha+n} G_0 + \frac{n}{\alpha+n} \frac{\sum_{i=1}^{n} \delta_{X_i}}{n} \qquad (3.34)$$

Therefore,

$$E(G \mid X_1, X_2, \ldots, X_n) = G_0^* \qquad (3.35)$$

As mentioned previously, like the Dirichlet distribution, with the change in the precision parameter of the Dirichlet process, the posterior distribution of $G \mid X_1, X_2, \ldots, X_n$ will tend to the prior distribution G_0 or the empirical distribution of the observations. G_0^* is the posterior base distribution that comes from a linear composition of the prior base distribution, G_0, and the empirical distribution of the observation $\frac{\sum_{i=1}^{n} \delta_{X_i}}{n}$.

That means if α grows, then $G_0^* \to G_0$, and if α decreases, then G_0^* trends to the distribution of observations. Additionally, when $n \to \infty$, G_0^* tends toward its limit distribution, which suggests that the Kolmogorov consistency theorem is also in place.

This result is the most fundamental feature of the Dirichlet process, which has given rise to a link between the process with the Pólya urn scheme, the Chinese

Dirichlet Process

restaurant process (CRP), and the Blackwell–MacQueen model, which are crucial in predicting the next steps of a stochastic process.

3.3 PÓLYA'S URN MODEL

Pólya's urn model is one of the specific states of urn models. The ideas in this model are made based on one or more bags/urns containing different colored balls inside them. Experiments by removing balls from the bag (with or without placement) lead to results that are widely used in discovering statistical models (Kottas, 2006; Tyoskin and Krivolapov, 1996).

For example, it may sometimes be important for the researcher to find the distribution of the number of balls of different colors, or for another researcher the distribution of the waiting time to reach a certain number of colored (or non-colored) balls. As mentioned, one of the urn models, which is also related to the Dirichlet distribution, is Polya's urn model. This model is often known as the Polya urn process or the Polya urn scheme.

3.3.1 Pólya's Urn Process

Consider an urn (or a bag) containing α balls including αm_j balls of each color for $j = 1, 2, \ldots, n$ and $0 < m_j < 1$. Suppose αm_j is an integer number. First, we randomly pull out a ball from the bag. Note the color of the ball and return it into the bag with another ball of the same color. Perform this test as many steps as possible. The goal is to calculate the probability of observation of jth color in the ith step. That means if X_i is the variable related to the color of the selected ball in the ith step, what is the answer to $P(X_i = j)$?

The answer to this probability varies from step to step. The general answer to this question is as follows:

$$P(X_1 = j) = \frac{\alpha m_j}{\sum_{i=1}^{n} \alpha m_i} = m_j$$

$$P(X_2 = j \mid X_1) = \frac{\alpha m_j + \delta(X_1 = j)}{\sum_{i=1}^{n} \alpha m_i + 1}$$

$$\vdots$$

$$P(X_{N+1} = j \mid X_{1:N}) = \frac{\alpha m_j + \sum_{i=1}^{N} \delta(X_i = j)}{\sum_{i=1}^{n} \alpha m_i + N}$$

(3.36)

This equation that is similar to equation (3.25) refers to the similarity of Polya's model to the Dirichlet process. As it is clear from the type of probabilistic structure of equation (3.36), this relation implies the predictive property of this equation given previous observations.

To generalize Polya's urn model, Blackwell and MacQueen proposed another design called the Blackwell–MacQueen urn scheme. To see more properties of this model, especially its predictivity, study the following sections.

3.3.2 BLACKWELL–MACQUEEN URN SCHEME

This scheme is a composition of Polya's model and a prior distribution. Suppose χ is the space of various colors and X_i is a variable related to the color of the ball picked in the ith step. First, take a color with a probability of G_0 (i.e., $X_i \sim G_0$) from the set of colors, χ, and smear a ball to that color and drop it into an empty bag/urn. In the second step, extract a new color with probability $\frac{\alpha}{\alpha+1}$ from the set of colors, χ, with distribution G_0 and/or remove a repetitive color with probability $\frac{1}{\alpha+1}$ from the set of colors extracted in the previous steps. In fact, in the second step, a new color with probability $\frac{\alpha}{\alpha+1}$ comes from distribution G_0 and a repetitive color with probability $\frac{1}{\alpha+1}$ comes from the previously selected colors. Note that in each step only one ball with a random color will be added to the bag.

For the purpose of the generalization of this process to reach the $(n+1)$th step, the $(n+1)$th color with probability $\frac{\alpha}{\alpha+n}$ shall randomly be selected from the set of colors by applying distribution G_0 and/or with probability $\frac{n}{\alpha+n}$ from the n colors selected in the previous steps.

Before formulating the proposed scheme, we present the definition that Blackwell and MacQueen have given about the Polya sequence.

Definition 3.3.1

The sequence $\{X_n, n \geq 1\}$ of random variables is called a Polya sequence with non-negative and finite parameter α, if for any subset $\beta \in \chi$:

$$P(X_1 \in B) = \frac{\alpha(B)}{\alpha(\chi)},$$

$$P(X_{n+1} \in B \mid X_1, X_2, \ldots, X_n) = \frac{\alpha_n(B)}{\alpha_n(\chi)}$$

(3.37)

where $\alpha_n = \alpha + \sum_{i=1}^{n} \delta(X_i)$.

To formulate the Blackwell–MacQueen urn scheme, suppose $X_1, X_2, \ldots, X_n$ is a vector of variables coming from the Dirichlet process G with precision parameter α,

Dirichlet Process

$$X_i \mid G \sim G, \quad i = 1, 2, \ldots, n$$

$$P(X_1, X_2, \ldots, X_n \mid G) = \prod_{i=1}^{n} P(X_i \mid G) \tag{3.38}$$

Note that the main goal is to formulate the predictor equation $P(X_{n+1} \mid X_1, X_2, \ldots, X_n, G)$.

With respect to Theorem 3.2.3, equations (3.25) and (3.36), and Definition 3.3.1,

$$X_1 \mid G \sim \mathrm{DP}(\alpha)$$

$$X_2 \mid X_1, G \sim \mathrm{DP}(\alpha + \delta(X_1))$$

$$X_3 \mid X_2, X_1, G \sim \mathrm{DP}(\alpha + \delta(X_1) + \delta(X_2))$$

Likewise for the $(n+1)$th step,

$$X_{n+1} \mid X_1, X_2, \ldots, X_n, G \sim \mathrm{DP}\left(\alpha + \sum_{i=1}^{n} \delta(X_i)\right) \tag{3.39}$$

According to Theorem 3.2.2, if X is a unit sample from the Dirichlet process G with precision parameter α, then

$$P(X \in A) = E(G(A)) = \frac{\alpha(A)}{\alpha(\chi)} \tag{3.40}$$

where $A \in \beta$. Now, let's stick with our running model (Blackwell–MacQueen model). By applying Theorem 3.2.2, the following results will come out:

$$P(X_{n+1} \in A \mid X_1, X_2, \ldots, X_n, G) = E(G(A) \mid X_1, X_2, \ldots, X_n, G)$$

$$= \frac{\alpha(A) + \sum_{i=1}^{n} \delta_{X_i}(A)}{\alpha(\chi) + n} \tag{3.41}$$

If αG_0 replaced α, equation (3.41) will be updated as follows:

$$P(X_{n+1} \in A \mid X_1, X_2, \ldots, X_n, G) = E(G(A) \mid X_1, X_2, \ldots, X_n, G)$$

$$= \frac{\alpha}{\alpha + n} G_0(A) + \frac{n}{\alpha + n} \frac{\sum_{i=1}^{n} \delta_{X_i}(A)}{n} \tag{3.42}$$

This equation shows the same results gained from equation (3.34). In fact, the base distribution of the posterior distribution, on condition of n previous observations, in the Dirichlet process is the predictor function of the $(n+1)$th observation.

This model, which is frequently referred to in the present section, is known as the Blackwell–MacQueen urn scheme which is applicable in simulating the data from the posterior distribution of the unknown parameters in the nonparametric Bayesian method.

Blackwell and MacQueen (1973) specified the relationship between Polya's sequence and the Dirichlet process by the presentation of the following theorem.

Theorem 3.3.1

Suppose $\{X_n, n \geq 1\}$ is a Polya sequence with parameter α. Then,

1. $\dfrac{\alpha}{\alpha_n(\chi)}$ leans toward a discrete measure, G, with probability one, when $n \to \infty$.
2. G is a Dirichlet process with precision parameter α.
3. Variables $X_1, X_2, \ldots$ given G are independent and have distribution G, that is,

$$X_1, X_2, \ldots | G \sim G$$

The next section presents one of the most important properties of the Dirichlet process.

3.4 DIRICHLET PROCESS AND CLUSTERING ISSUE

The Dirichlet process also has the data clustering property because of the dissociation property of the data. As can be seen from relation (2.10), with probabilities $\dfrac{n}{\alpha+n}$ and $\dfrac{\alpha}{\alpha+n}$, this process generates repetitive data from the previous observations and new data from the base distribution G_0, respectively.

Note that clusters are sets of duplicate data that may be generated using equation (3.34) and according to the Blackwell–MacQueen model. For example, according to the Blackwell–MacQueen scheme, the total number of balls of the same color can be considered a cluster. Hence, any repetitive data (or, equivalently, a ball with the same color) is included within a set, which is called a cluster.

For instance, during a simulation process using equation (3.34), it may n data, including new observations or repetitive data, get generated wherein x_1 repeats n_1 times, x_2 repeats n_2 times, and finally x_{n^*} repeats n_{n^*} times in which $n = \sum_{i=1}^{n^*} n_i$.

The property of generating duplicate data, which is due to the discrete feature of the Dirichlet process in the extraction of new data, creates distinct clusters between the data. To get more understanding of this concept, consider the following example:

Suppose $X_1, X_2, \ldots, X_n$ are variables of the Dirichlet process with parameter αG_0. To get new observations of this process using the predictive equation of the Blackwell–MacQueen, suppose X_i refers to the color of the ith ball for $i = 1, 2, \ldots, n$. If n_j, for $j = 1, 2, \ldots, k$, is the number of balls with the color j, $X_1^*, X_2^*, \ldots, X_k^*$ is a

Dirichlet Process

sample of the Dirichlet process that X_j^* refers to the variable showing the color of the jth ball. Clearly, each X_j^* creates a distinct cluster. In another word, the sample of size n is reduced to a sample of size k (i.e. $k \leq n$). Additionally n_j, for $j = 1, 2, \ldots, k$, is the number of repetitive data in jth cluster. Predictive equation (3.42) can be rewritten as follows:

$$P(X_{n+1}|X_1, X_2, \ldots, X_n)$$

$$= \frac{\alpha}{\alpha+n} G_0 + \frac{1}{\alpha+n} \sum_{j=1}^{k} n_j \delta(X_j^*) \qquad (3.43)$$

According to equation (3.43), the new data of variable X_{n+1} is generated from G_0 with probability $\frac{\alpha}{\alpha+n}$ and is generated from previous data, proportional to the jth cluster, with probability $\frac{1}{\alpha+n}$.

Note that the larger the subsample size n_j, the greater the likelihood that $(n+1)$th new data will be selected from the jth cluster. In addition, the $(n+1)$th new data will be generated from G_0 when $\alpha \to \infty$.

In fact, when $\alpha \to \infty$, all data will come from G_0 that refers to a non-repetitive sample in which the number of clusters is equal to the sample size n (and vice versa). To illustrate this, see the process of clustering the data generated from the Dirichlet process, using the Blackwell–MacQueen model, in the following example.

Example 3.4.1

This example illustrates how the number of clusters changes when the precision parameter α varies. The base distribution is considered to be a beta distribution with known parameters $(\alpha, \beta) = (2,3)$. That means $G_0 = \text{Beta}(2,3)$. As Figure 3.6 demonstrates, the number of clusters increases when the precision parameter α grows (and vice versa). The size of the sample generated from the present Dirichlet process with beta base distribution and various amounts of the precision parameter is 31.

If the vertical axis shows the number of clusters and the horizontal axis shows the location of each unit of the sample, the interpretation of Figure 3.6 in order of clockwise from top-left is as follows.

The first graph, the top-left figure, relates to the number of clusters made by 31 units which have been generated from the Dirichlet process with the known beta base distribution and precision parameter $\alpha = 0.1$. As this figure shows, the number of clusters made by this sample is 1. The main reason causing this result is mostly due to the low amount of the precision parameter. In fact, because of the minor value of this parameter in this specific example, the new sample units generated from this process are totally repetitive.

The second graph pertains to the number of clusters created by the 31 sample units generated by this process with precision parameter $\alpha = 1$. Because of the greater amount of α in this example, as compared to the first graph, the number of clusters in this example has increased to 2. The third and fourth graphs show the

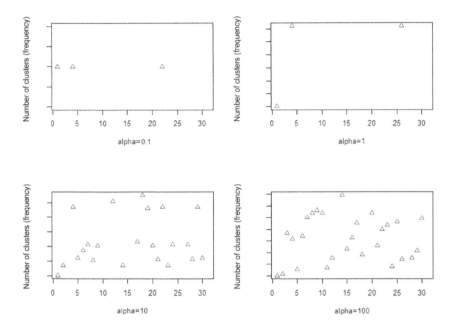

FIGURE 3.6 Number of the clusters created by the data generated from the Dirichlet process with known base distribution, Beta(2,3), and different amounts of the precision parameter ($\alpha(0.1,1,10,100)$). (Authors' own figure.)

number of the clusters made by the samples generated from the Dirichlet process with precision parameters 10 and 100, respectively.

Naturally, the number of clusters will grow as compared to the previous graphs. Therefore, the number of clusters in the fourth graph, which is drawn for $\alpha = 100$, is almost equal to the number of sample units. In fact, because of the high amount of the precision parameter in the last graph, each new unit generated from this process makes a new cluster.

To use the programming codes of Figure 3.6, see Section B.9 in the Appendix.

3.4.1 CHINESE RESTAURANT PROCESS

One of the important issues in the Dirichlet process is the determination of the number of clusters (partitions) generated by the data simulated by this process. In order to obtain the mean and variance of the number of clusters, it is necessary to construct the probability function of the partitions obtained in the Dirichlet process. Hence, because of its capability, the CRP, which gets its name from a Chinese restaurant based in San Francisco, has been considered to determine the number of clusters.

To illustrate the structure of the CRP, imagine a restaurant containing infinite tables in which each table contains infinite chairs. The first customer selects one of the tables with probability 1. The second customer decides to sit at an empty table or the table occupied by the previous customer. With a probability proportional to α, the

Dirichlet Process

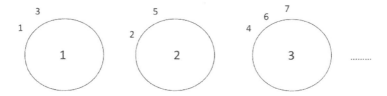

FIGURE 3.7 An example of the number of clusters created by seven customers when entering a restaurant. (Authors' own figure.)

new customer decides to choose a non-occupied table or, with a probability proportional to 1, the new customer decides to select a chair around the table that has been occupied by the previous customer. To generalize this issue to n customers, suppose n_i is the number of customers around the ith table.

Hence, the nth customer selects a new table proportional to α and/or select an occupied table with a probability proportional to n_i. To get more understanding, consider the following example.

According to Figure 3.7, suppose seven customers enter a restaurant containing infinite tables with infinite chairs. Imagine the seating arrangement for these seven customers as shown in Figure 3.7.

Suppose X_i is the variable related to the location (or chair number) of the ith customer. Therefore, by applying this variable, the probabilities of location (or chair number), of the ith customer will be calculable.

Consider the following result:

$$P(X_1, X_2, \ldots, X_7) = P(X_1) P(X_2 \mid X_1) \ldots P(X_7 \mid X_1, X_2, \ldots, X_6)$$
$$= \left(\frac{\alpha}{\alpha}\right)\left(\frac{\alpha}{1+\alpha}\right)\left(\frac{1}{2+\alpha}\right)\left(\frac{\alpha}{3+\alpha}\right)\left(\frac{1}{4+\alpha}\right)\left(\frac{1}{5+\alpha}\right)\left(\frac{2}{6+\alpha}\right) \quad (3.44)$$

As can be seen from the above probability relation, there will be no change in the probability value if the sixth customer order shifts with the seventh customer. Therefore, in this probability equation, there is no order. That is, in the CRP, a new customer who enters the restaurant can be considered as the last person who chooses a new or occupied table with the proportional likelihood.

Since in the CRP each table chosen by every customer happens completely randomly, this causes the number of clusters to be randomized in this process. Also, according to the finite exchangeable sequences and De Finetti's theorem that is true in the process, there is no difference in customer displacement. Therefore, this will play a key role in the simulation algorithm for generating new data and determining the number of clusters.

Suppose ϑ_i, for $i = 1, 2, \ldots, n$, is an indicator variable pertaining to the ith customer so that $\vartheta_i = 1$ if the ith customer chooses a new table, and $\vartheta_i = 0$ if he/she decides to take a seat around an occupied table. If K is the variable related to the number of occupied tables (or, equivalently, the number of clusters) created by n customers, then the expected value and the variance of the number of clusters can be calculated as follows:

$$E(K|n) = E\left(\sum_{i=1}^{n} \vartheta_i \mid n\right) = \sum_{i=1}^{n} P(\vartheta_i = 1) = \sum_{i=1}^{n} \frac{\alpha}{\alpha + i - 1}$$
$$= \alpha(\psi(\alpha + n) - \psi(\alpha)) \approx \alpha \text{Log}\left(1 + \frac{n}{\alpha}\right)$$
(3.45)

and

$$\text{Var}(K|n) = \alpha(\psi(\alpha+n) - \psi(\alpha)) + \alpha^2\left(\psi(\alpha+n)' - \psi(\alpha)'\right)$$
$$\approx \alpha\left(\text{Log}\left(1 + \frac{n}{\alpha}\right) - 1\right)$$
(3.46)

where $\psi(.)$ is the Digamma function. According to the equation obtained for the mean and variance of the number of clusters in this process, it is obvious that with increasing α value the number of clusters also increases and with the increase in the number of customers, n, the number of clusters will grow logarithmically. The reason for the gradual increase in the number of clusters with respect to n, as compared to α, is due to the increase in the number of customers within each cluster. If the chance for a new customer who enters a restaurant to choose the ith table repetitively is proportional to n_i, the likelihood of another new customer to select the same table grows if n_i increases, and then the trend of increasing the number of clusters with respect to n is slower compared to α.

To see the relationship between the number of clusters and the precision parameter in the CRP, keep reading with Example 3.4.2.

Example 3.4.2

Suppose 40 customers, $n = 40$, enter a restaurant one after another. The process of changing the number of clusters by these 40 customers for different values of the precision parameter can be seen in Figure 3.8. For $\alpha = 0.1$, only one cluster is formed during the entry of 40 customers, that is, all customers are seated around a table. For $\alpha = 1$, the customers are seated around two distinct tables. For $\alpha = 10$, there are 17 clusters, and for $\alpha = 100$, there are about 35 distinct clusters. To use the programming codes of Figure 3.8, see Section B.10 in the Appendix.

These graphs illustrate the fact that as the value of α increases, customers are more likely to choose a new table. This will greatly help with the Dirichlet process simulation problem to obtain new values. Since the existence of duplicate data does not add any new information to our sample set, it is attempted to use this process to generate new and non-duplicate samples in the next chapter.

In Chapter 2, we briefly examined the Dirichlet distribution and the Dirichlet process and some of their features. After understanding this process and how it behaves for different parameters, in the second part, we studied the process of Polya's urn and its relation to the Dirichlet process. On the other hand, we introduced a prediction method using the Blackwell–MacQueen urn scheme. Finally,

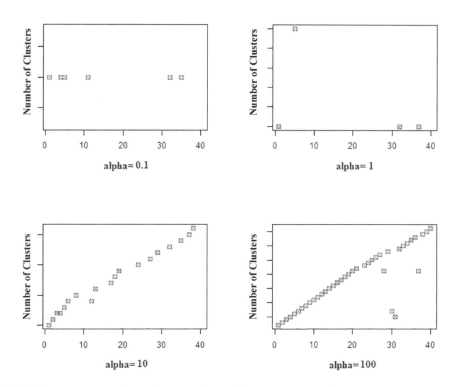

FIGURE 3.8 The number of clusters in the CRP when the precision parameter changes. (Authors' own figure.)

we analyzed one of the features of the Dirichlet process, namely data clustering, using the CRP.

In Chapter 4, we first review mixture models and then introduce a regression model called the semi-parametric log-linear model, which is used in accelerated lifetime tests. In the next section, we briefly study the methods of determining the base distribution and the precision parameter of the Dirichlet process. We then analyze the Dirichlet process mixture model with a kernel. Finally, we present the simulation algorithm for estimating unknown parameters using the Markov chain Monte Carlo (MCMC) method.

4 Nonparametric Bayesian Approach in Accelerated Lifetime Tests

The advancement of technology in today's world has led to the production of high-quality products. Therefore, researchers need a lot of time and cost to analyze the lifetime of products at normal stress levels, and this has led researchers to adopt accelerated lifetime tests (ALTs). Because in ALTs, products are at a higher stress level than normal, so products' lifetime reduces and this will reduce the test costs.

The accelerated lifetimes obtained under these tests by one of the stress functions (which are referred to in Chapter 2) convert the product lifetime to the normal stress level, and finally, the data obtained from these convert functions will be used to assess the reliability function, hazard rate, probability density function (PDF), cumulative distribution function (CDF), and other features.

According to Nelson (1990) and Meeker and Escobar (1998), ALTs require two important hypotheses.

The **first assumption** is about the failure-time distribution. Under this assumption, elements under all stress levels shall have similar time distribution. Since the failure time is equivalent to the lifetime data, the families of location-scale distributions such as exponential, Weibull, and log-normal can be used as the failure-time distribution.

The **second assumption** relates to the relationship between the parameters of the fracture-time distribution and stress variables using an appropriate acceleration function. Since parametric models place limiting assumptions on the data distribution (and in Bayesian analysis limiting assumptions on the parameters), due to the flexibility of the nonparametric models, these models are taken into consideration by researchers and scholars.

Tyoskin and Krivolapov (1996), Basu and Ebrahimi (1982), Shaked and Singpurwalla (1982), Lin and Fei (1991), Bai and Chun (1993), and Bai and Lee (1996) have published papers on nonparametric analysis of ALTs, and there is no limiting assumption on the distribution of data and parameters in their methods.

The method used here is also a nonparametric one that will be associated with the Bayesian approach. Therefore, according to a nonparametric analysis from a Bayesian perspective, it is necessary to use a family of distributions that is noninformative, rather than a specific prior distribution for parameters which is a limiting assumption.

Therefore, according to the content mentioned throughout Chapter 3, the Dirichlet process can be a solution to such problems.

In this chapter, we first examine mixture models and then introduce a regression model called the semi-parametric log-linear regression, which is used in ALTs. In the next section, we briefly study the methods of determining the base distribution and the precision parameter of the Dirichlet process. We then analyze the Dirichlet process mixture model with a kernel. Finally, we present the simulation algorithm for estimating unknown parameters using the Markov chain Monte Carlo (MCMC) method.

4.1 DIRICHLET PROCESS MIXTURE MODELS

To examine the Dirichlet process mixture models, foremost, take notice of the brief description of the following mixture models.

4.1.1 Mixture Models

Sometimes in the statistical analysis, we may come across data that contain complex shapes so that certain distributions cannot be applied. For example, considering a simple and unimodal distribution on a complex dataset with a multimodal histogram graph will impose a great error in results.

Suppose $\vartheta_1, \ldots, \vartheta_n$ is a sample of an unknown family $f(\vartheta)$. A mixture probability density function, with M components, is defined as follows:

$$f(\vartheta | \pi_1, \ldots, \pi_M, \theta_1, \ldots, \theta_M) = \sum_{j=1}^{M} \pi_j k_j(\vartheta | \theta_j) \tag{4.1}$$

where $k(\vartheta | \theta)$ is a parametric kernel with parameter vector $\theta \equiv (\theta_1, \ldots, \theta_M)$. In addition, $\pi \equiv (\pi_1, \ldots, \pi_M)$ is known as weight parameter in which, for $j = 1, \ldots, M$, $0 < \pi_j < 1$ and $\sum_{j=1}^{M} \pi_j = 1$.

As this equation shows, the density function $f(\vartheta)$ is made by M different density functions. In Bayesian inference, a suitable distribution can be fitted to the data by selecting the appropriate prior for the parameters in the mixture models.

Figure 4.1 is an example of functions related to the simulated data from a mixture model. These four diagrams show that these data are derived from a combination of two different populations, with 40% of the first population (namely $\pi_1 = 0.4$) and 60% of the second population (namely $\pi_2 = 0.6$). Therefore, to study this dataset, it is better to use a mixture of two probability density functions (PDFs).

To use the programming codes of Figure 4.1, see Section B.11 in the Appendix.

A fundamental issue in mixture models is the correct detection of the density of ith sample unit (ϑ_i). For this purpose, a latent variable like c_i, which is related to ϑ_i, is defined for indicating that ϑ_i comes from which one of the M density functions (components).

Görür and Rasmussen (2010) have introduced a hierarchical form for parametric mixture models as follows:

Nonparametric Bayesian Approach in ALTs

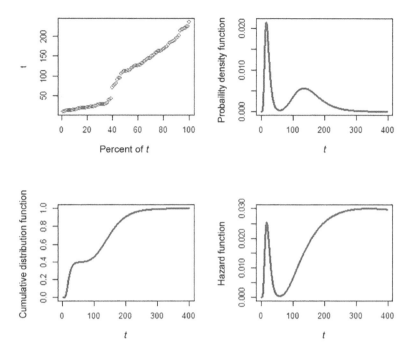

FIGURE 4.1 Scatter plot, PDF, CDF, and hazard function of the data obtained from two different log-normal populations. (Authors' own figure.)

$$\vartheta_i \mid c_i \sim k(\vartheta_i \mid \theta_{c_i}), \quad i = 1, 2, \ldots, n$$

$$\theta_j \sim G_0(\theta), \quad j = 1, 2, \ldots, M$$

$$c_i \sim \text{Discrete}(\pi_1, \ldots, \pi_M), \quad i = 1, 2, \ldots, n \quad (4.2)$$

$$\pi \mid \alpha \sim \text{Dir}\left(\frac{\alpha}{M}, \ldots, \frac{\alpha}{M}\right)$$

where α is the precision parameter of the Dirichlet distribution and $G_0(\theta)$ is the prior distribution of the parameter θ_j for $j = 1, 2, \ldots, M$. Additionally, c_i, for $i = 1, 2, \ldots, n$, has a discrete distribution as follows:

$$p(c_1, \ldots, c_n \mid \pi) \propto \prod_{j=1}^{M} \pi_j^{n_j} \quad (4.3)$$

where n_j is the number of observations belonging to component j. Hence, $\sum_{j=1}^{M} n_j = n$. According to Görür and Rasmussen (2010), the distribution of $n, \ldots, n_M$ is a multinomial as follows:

$$p(n_1, \ldots, n_M | \pi) = \frac{n!}{n_1! n_2! \ldots n_M!} \prod_{j=1}^{M} \pi_j^{n_j} \tag{4.4}$$

Since in the hierarchical model (3.1), the number of parameters is finite, in Bayesian inference, a suitable prior for each unknown parameter is considered and, using Bayes' law, the posterior distribution and, finally, the Bayesian estimation of the parameters can be calculated. Further studies on mixture models can be found in the articles published by Müller and Quintana (2004), MacEachern and Müller (1998), and Escobar and West (1995).

Now, if in the mixture model (3.1) $M \to \infty$, namely if the number of components of the mixture model becomes infinite, then the number of unknown parameters in this model will also be infinite. Therefore, the Bayesian inference that will be used in this case would be based on the nonparametric approaches. It should be noted, as mentioned in the previous chapter, that the difference between the parametric Bayes and nonparametric Bayes is the finite and infinite numbers of unknown parameters, so, in this case, the prior distribution for the parameters is considered to be a stochastic process. Since the Dirichlet process is used in models with infinite unknown parameters, and it has striking features, this stochastic process can be considered as the prior distribution of parameters in the mixture models with infinite components (Ferguson, 1973; Singpurwalla, 2006).

Therefore, in the nonparametric state, the hierarchical form (3.1) of the mixture models can be rewritten into the hierarchical form of the following Dirichlet process:

$$\vartheta_i | \theta_i \sim k(\vartheta_i | \theta_i), \quad i = 1, 2, \ldots, n$$
$$\theta_i | G \sim G(\theta), \quad i = 1, 2, \ldots, n \tag{4.5}$$
$$G(\theta) \sim \mathrm{DP}(\alpha, G_0(\theta))$$

Note that, in the nonparametric approach, $G_0(.)$ refers to the base distribution of the Dirichlet process. According to the above two mixture models, i.e., finite and infinite, the main difference between the parametric mixture model and the nonparametric mixture model, also known as the Dirichlet process mixed model, is in the number of components or equivalently in the number of parameters. Also in the Dirichlet process mixture model, all parametric vectors, θ_i, are extracted from the distribution $G(\theta)$, while in the parametric mixture model, these vectors are derived from the prior distribution $G_0(\theta)$.

The Dirichlet process mixture model is a nonparametric Bayesian data analysis tool that is widely used in estimating nonparametric density and regression models as well as is applied in survival and reliability data analysis (Escobar and West, 1995; Müller et al., 1996; Kottas, 2006).

The unknown density function $f(\vartheta)$, which is modeled by the Dielectric process mixture model, can be represented by the marginal density as follows:

$$f(\vartheta | G) = \int k(\vartheta | \theta) dG(\theta) \tag{4.6}$$

Since we need to simulate the parameters of the Dielectric process to predict the marginal function in (2.3), and also since in the Dirichlet process mixture model we are faced with a large number of parameters, to estimate parameters by applying Bayes' rule, under the desired loss function, it is hard to obtain the posterior density function. Hence, the method for simulating these parameters is the MCMC simulation algorithm.

According to Ferguson's first assumption in lifetime tests, it is necessary to establish a relationship between model parameters and different stress levels. To express this relationship, a model called the log-linear regression model is used, which is defined in the following subsection.

4.2 LOG-LINEAR REGRESSION IN THE NONPARAMETRIC PROBLEM

According to Kuo and Mallick (1997), a semiparametric regression model is used to infer the nonparametric Bayes in ALTs. The regression model used in inference related to lifetime tests is considered to be log-linear regression. The coefficients in this regression model are estimated by applying Bayesian or classic methods. For instance, in the Bayesian method, a prior distribution is considered for unknown coefficients, and by using the Bayesian law, the posterior distribution of these parameters is obtained on the condition of data. Finally, using the desired loss function, the unknown coefficients can be estimated using the calculated posterior density function.

Note that, according to Ghosh et al. (2007), these unknown coefficients can also be calculated by applying the maximum likelihood method which is a classic and frequency approach.

Suppose $T = (T_1, T_2, \ldots, T_n)$ is a vector of variables related to the lifetime of n products. The log-linear regression model for $i = 1, 2, \ldots, n$ is as follows:

$$\text{Log } T_i = -x_i' \beta + W_i \tag{4.7}$$

where $x_i = (x_{i1}, x_{i2}, \ldots, x_{ip})$ is the regression variables related to the ith product and $\beta_{p \times 1}$ is the vector of unknown parameters related to the regression coefficients. W_i is also a regression error that is usually considered independent and identically distributed that comes from a parametric family, such as normal, logistic, gamma, and Weibull distributions (Kalbfleisch and Prentice, 1980).

In the log-line regression model, in addition to the parametric part, which is related to the regression coefficients, there is also a nonparametric part, which is an unknown error. Since using a parametric distribution to estimate errors will impose a restrictive assumption on the model, it is tried to apply this nonparametric method to fit the distribution of the errors, too. In the nonparametric regression model, a method composed of parametric and nonparametric methods, also named semiparametric methods, will be used.

Since the nonparametric Bayes method requires a prior distribution (and given the properties of the Dirichlet process described in the previous chapter), the Dirichlet process can also be used as a prior distribution to log-line regression errors. Because the Dirichlet process mixture model is more flexible than the Dirichlet process, this model, along with a continuous kernel, can be used as a prior for errors in the log-linear regression model.

Equation (5.35) can be rewritten as follows (Christensen and Johnson, 1988):

$$T_i = e^{-x_i'\beta + W_i} = e^{-x_i'\beta} V_i \tag{4.8}$$

where $V_i = e^{W_i}$. Since V_i takes positive values, the distribution for this variable can be Weibull, gamma, and log-normal or any other distribution with a positive domain. The value of W_i can also be negative, so distributions with a negative domain such as normal distribution can also be considered as candidates to fit these errors.

Suppose V_i, for $i = 1, 2, \ldots, n$, are independent random variables and identically distributed as follows:

$$f(v_i | G) = \int k(v_i | \theta_{1i}, \ldots, \theta_{li}) G(d\theta_{1i}, \ldots, d\theta_{li}) \tag{4.9}$$

where l is the number of unknown parameters in the kernel $k(v_i | \theta_{1i}, \ldots, \theta_{li})$. In addition, $\Theta_i = (\theta_{1i}, \ldots, \theta_{li})$ for $i = 1, 2, \ldots, n$ makes a matrix of unknown parameters in which each column is related to the ith observation.

As a specific example, suppose $k(v_i | \theta_{1i}, \ldots, \theta_{li})$ refers to a two-parameter Weibull distribution namely Weibull (α, λ). So $l = 2$ and, for $i = 1, 2, \ldots, n$, $\theta_{1i} = \alpha_i$ and $\theta_{2i} = \lambda_i$. Additionally, the distribution function and the density function of this distribution are as follows:

$$k(v | \alpha, \lambda) = \lambda^{-1} \alpha v^{\alpha - 1} e^{-\lambda^{-1} v^\alpha}$$
$$K(v | \alpha, \lambda) = 1 - e^{-\lambda^{-1} v^\alpha}, \quad \alpha > 0, \lambda > 0, v \geq 0 \tag{4.10}$$

By applying the Jacobi method, the marginal density function of variable T_i can be obtained as follows:

$$f(t_i | \beta, G) = f_{v_i}\left(t_i e^{x_i'\beta} | G\right) \left|\frac{dv_i}{dt_i}\right|$$
$$= \int \left|\frac{dv_i}{dt_i}\right| k\left(t_i e^{x_i'\beta} | \theta_{1i}, \ldots, \theta_{li}\right) G(d\theta_{1i}, \ldots, d\theta_{li}) \tag{4.11}$$

where $\left|\frac{dv_i}{dt_i}\right|$ is the absolute value of the Jacobi transformation. Likewise, $k(. | \theta_{1i}, \ldots, \theta_{li})$ is the kernel of the given density function.

By considering the Weibull distribution as a specific example, the cumulative distribution function of variable T_i is achievable as follows:

$$F(t_i|\beta, G) = \Pr(T_i \leq t_i|\beta, G)$$

$$= \int_0^{t_i} f(s|\beta, G)\,ds$$

$$= \int\int_0^{t_i} e^{x_i'\beta} k\left(se^{x_i'\beta}|\alpha_i, \lambda_i\right) G(d\alpha_i, d\lambda_i)\,ds \quad (4.12)$$

$$= \int \int_0^{t_i} e^{x_i'\beta} k\left(se^{x_i'\beta}|\alpha_i, \lambda_i\right) ds\, G(d\alpha_i, d\lambda_i)$$

$$= \int K\left(t_i e^{x_i'\beta}|\alpha_i, \lambda_i\right) G(d\alpha_i, d\lambda_i)$$

It is easy to show that if $V_i \sim$ Weibull, then variable T_i also has the Weibull distribution. The density function of the variable T_i is as follows:

$$f(t_i|\alpha_i, \lambda_i) = e^{\alpha_i x_i'\beta} \lambda_i^{-1} \alpha_i t_i^{\alpha_i-1} e^{-e^{\alpha_i x_i'\beta}\lambda_i^{-1} t_i^{\alpha_i}}; \quad \alpha_i > 0, \lambda_i > 0, t_i \geq 0 \quad (4.13)$$

Now if define $\theta_i^{-1} = e^{\alpha_i x_i'\beta} \lambda_i^{-1}$, then

$$f(t_i|\alpha_i, \lambda_i) = \theta_i^{-1} \alpha_i t_i^{\alpha_i-1} e^{-\theta_i^{-1} t_i^{\alpha_i}}; \quad \alpha_i > 0, \theta_i > 0, t_i \geq 0 \quad (4.14)$$

4.3 DETERMINING THE BASE DISTRIBUTION AND THE PRECISION PARAMETER

The base distribution, G_0, plays a very important role in the data analysis results in the Dirichlet process mixture models, and also the choice of base distribution is much influenced by the kernel distribution that is adopted in the mixture model. Often, in order to avoid difficult and complex calculations, in obtaining conditional posterior distributions using the Dirichlet process prior, the base distribution is considered to be conjugate. In general, the Gaussian Dirichlet process mixture model has wide applications in both conjugate and nonconjugate base distributions (Escobar, 1994).

As stated in Chapter 3, the precision parameter in the Dirichlet process controls the proximity of the process to the base distribution. As Antoniak (1974) pointed out, the magnitude of the precision parameter indicates the degree of confidence in the base distribution, and this parameter can play a key role in modeling as an effective factor in the number of clusters around the base distribution.

Likewise, Görür and Rasmussen (2010) had a similar idea. They showed that for the concentration parameter (alternatively the precision parameter) a noninformative prior can be considered in which the estimated values of the precision parameter can be obtained by applying Bayes' law.

Liu (1996) defined the precision parameter as a definite value that falls into the interval [0,1]. According to this structure, $c.\text{Beta}(a_\mu, b_\mu)$ is considered as the prior distribution for the precision parameter in which $\text{Beta}(a_\mu, b_\mu)$ refers to the beta distribution with unknown/known parameters. Additionally, c is considered as the weight parameter that takes positive values. Antoniak (1974), Beckett and Dlaconis (1994), and Korwar and Hollander (1973) worked on c and proved that this parameter affects the number of clusters made by simulated data. According to their findings, the big amount of c smoothens the conditional posterior distribution of the precision parameter.

Therefore, according to studies, for ease of obtaining conditional posterior distributions, researchers consider the base distribution of the Dirichlet process to be usually a conjugate to the applied kernel. In addition, this will facilitate not only the calculation of the conditional posterior distribution but also the simulation steps. Generally, in order to increase accuracy as well as not to impose restrictions on the precision parameter, a prior distribution with a nonnegative range for this parameter is usually considered.

4.4 HIERARCHICAL MODEL OF THE DIRICHLET PROCESS

Suppose $t_1, t_2, \ldots, t_n$ is the lifetime of n products and consider x_i shows the stress level that is imposed on the ith product. Additionally, suppose δ_i is an indicator variable showing the sensor or complete situation of the lifetime of the ith product, namely $\delta_i = 1$ if the lifetime of the ith product is complete and $\delta_i = 0$ if the data is censored.[1]

Given the earlier explanations, lifetime data will be demonstrated using the following notation:

$$d = \{(t_i, x_i, \delta_i), i = 1, 2, \ldots, n\} \quad (4.15)$$

In addition to the main parameters in the studied density, as the kernel of the mixture model, as well as the regression coefficient parameter, β, there may be another set of parameters according to the type of the base distribution of the Dirichlet process that is called hyperparameters.

In general, throughout this book, all hyperparameters and the precision parameter will be notated by the set $\xi = (\alpha, \xi_i)_{i \geq 0}$.

In the hierarchical Bayesian analysis, specific distributions may be considered as the distribution for each of the hyperparameters in several steps, where each of those considered distributions may have its unknown parameters. Since, in the hierarchical analysis, the effect of the last parameter on the main parameter decreases with the increasing number of steps, the last parameter is considered as a fixed value (provided that the number is valid in the domain of the parameter).

---

[1] Note that in order to have a step-by-step education on how to apply the nonparametric Bayesian approach, we used the accelerated lifetime dataset to support both nonparametric Bayes and reliability analysis. Additionally, the incomplete data used here is the censored lifetime data from the right side which is called the right-censored data.

Nonparametric Bayesian Approach in ALTs

With all the definitions and assumptions that have been considered so far, in general, the hierarchical form of the Dirichlet process mixture model with the desired kernel can be written as follows:

$$t_i \mid \theta_{1i}, \theta_{2i}, \ldots, \theta_{Ki}, \beta \sim \begin{cases} f(t_i \mid \theta_{1i}, \theta_{2i}, \ldots, \theta_{Ki}); & \text{if } \delta_i = 1, \quad i = 1, 2, \ldots, n \\ 1 - F(t_i \mid \theta_{1i}, \theta_{2i}, \ldots, \theta_{Ki}); & \text{if } \delta_i = 0 \end{cases}$$

$$(\theta_1, \theta_2, \ldots, \theta_K) \mid G \sim G(\theta_1, \theta_2, \ldots, \theta_K), \quad i = 1, 2, \ldots, n$$

$$G(\theta_1, \theta_2, \ldots, \theta_K) \sim \text{DP}(\alpha, G_0(\theta_1, \theta_2, \ldots, \theta_K)),$$

$$\xi_i \sim f_{\xi_i}(\xi_i \mid a_{\xi_{i1}}, \ldots, a_{\xi_{iL}}), \quad i \geq 1$$

$$\alpha \sim f_\alpha(\alpha \mid b_{\alpha_{i1}}, \ldots, b_{\alpha_{iM}}), \quad i \geq 1 \tag{4.16}$$

$$\beta \sim f_\beta(\beta \mid c_{\beta_{i1}}, \ldots, c_{\beta_{iN}}), \quad i \geq 1$$

where L, M, and N are the number of hyperparameters in the relevant prior density function of each unknown parameter that may vary from parameter to parameter. The main reason for the differentiation of the density function into complete and censored data is the use of the likelihood function of censored data to facilitate the way reaching the posterior density function of parameters.

4.5 BAYESIAN COMPUTATION

Calculating the posterior distribution for the high-dimensional functions in the Bayesian inference is one of the most difficult issues. For this reason, various simulation methods are used to generate data from the posterior distribution. Therefore, due to the capability and flexibility of the MCMC method in generating random data, this method is widely used (Walsh, 2004).

As mentioned in Chapter 2, the main difference between the MCMC method and other sampling methods is the presence of Markov chain properties in this method. This means that simulation at one stage always depends on the previous step. More importantly, of all the MCMC sampling methods, the Gibbs sampling method, where conditional distribution and the joint distribution can be sampled by it, is most applicable (Giudici et al., 2009).

Our goal in this section is to provide a general algorithm using the Gibbs sampling method to fit a mixture model by applying the Dirichlet process to the accelerated lifetime data. This algorithm will be exploited by applying numeric illustrative examples in Chapter 5.

Due to the unlimited number of parameters in the Dirichlet process mixture model, the Gibbs sampling method will work well. On the other hand, when the sample size is large enough, it can be assumed that the simulated values of a parameter with the Gibbs sampling method are extracted from the posterior distribution of the parameter, and so, this statistical sample can be used to estimate the parameter (Gelman et al., 2003).

Before expressing the general framework of the sampling algorithm, we rewrite the Blackwell–MacQueen urn scheme, equation (3.42), by applying the posterior distribution. As mentioned previously, equation (3.42) is capable to be used in predicting new observations. This equation can be rewritten with respect to the unknown parameter θ as follows:

$$\Pr(\theta_{n+1} = j | \theta_1, \ldots, \theta_n, \alpha, G_0) = \begin{cases} \dfrac{1}{\alpha+n}\sum_{i=1}^{n}\delta(\theta_i = j), & \exists k \leq n \quad \text{S.T.} \quad \theta_k = j \\ \dfrac{\alpha}{\alpha+n} G_0, & \forall\ 1 \leq k \leq n \quad \text{S.T.} \quad \theta_k \neq j \end{cases}$$

(4.17)

Now, by applying Bayes' law, the posterior distribution of θ_i, for $i = 1, 2, \ldots, n$, is calculated as follows:

$$h(\theta_i | \theta_{-i}, t_i) \propto f(t_i | \theta_i)\pi(\theta_i | \theta_{-i}) \qquad (4.18)$$

And the Blackwell–MacQueen model is

$$\Pr(\theta_i | \theta_{-i}, t_i) = bq_0 h(\theta_i | t_i) + b \sum_{\substack{j=1 \\ j \neq i}}^{n} f(t_i; \theta_j)\delta(\theta - \theta_j) \qquad (4.19)$$

where $h(.|t_i)$ is the posterior density function and b is the normalizing constant which are defined as follows:

$$h(\theta_i | t_i) = \dfrac{G_0(\theta_i) f(t_i; \theta_i)}{\int_\theta G_0 f(t_i | \theta)d\theta},$$

$$b = \left(q_0 + \sum_{\substack{j=1 \\ j \neq i}}^{n} f(t_i; \theta_j) \right)^{-1} \qquad (4.20)$$

$$q_0 = \alpha \int_\theta G_0 f(t_i | \theta) d\theta$$

To prove this relation, see Section A.5 in the Appendix.

Due to the clustering property of the Dirichlet process and the relationship between this stochastic process and the Blackwell–MacQueen urn scheme, we will follow the following probability function to extract the parameter θ_i:

$$Pr(\theta_i | \theta_{-i}, t_i) \begin{cases} = \theta_j & \text{with a probability of } bf(t_i; \theta_j) \\ \approx h(\theta | t_i) & \text{with a probability of } bq_0 \end{cases} \qquad (4.21)$$

4.6 MODEL FITTING

Suppose $\Theta = (\theta_{1i}, \theta_{2i}, \ldots, \theta_{ki})$, for $i = 1, 2, \ldots, n$, is the vector of unknown parameters in a Dirichlet process that would be used to study a set of accelerated lifetime data. The main goal in this subsection is to generate data from the conditional distribution below by applying the Gibbs sampling method:

$$f(\theta_{11}, \ldots, \theta_{1n}, \theta_{21} \ldots, \theta_{2n}, \theta_{k1} \ldots, \theta_{kn}, \beta, \xi | d) \qquad (4.22)$$

where $\xi = (\alpha, \xi_i)_{i \geq 0}$ encompasses the precision parameter and all hyperparameters. Besides, d refers to the vector of observations/data, which may be complete or censored or a combination of both.

If we define $\Theta_i = (\theta_{1i}, \theta_{2i}, \ldots, \theta_{ki})$ and $\Theta_j^* = (\theta_{1j}^*, \theta_{2j}^*, \ldots, \theta_{kj}^*)$, to generate data with the use of the Gibbs sampling method, the posterior distributions of the following variables are required:

$$\Theta_i | (\Theta_{-i}, c_{-i}), \xi, \beta, d \quad \text{for } i = 1, 2, \ldots, n,$$

$$\Theta_j^* | c, n^*, \xi, \beta, d \quad \text{for } j = 1, 2, \ldots, n^*,$$

$$\alpha | \{\Theta_j^*, j = 1, 2, \ldots, n^*\}, n^*, d, \qquad (4.23)$$

$$\xi_i | \{\Theta_j^*, j = 1, 2, \ldots, n^*\}, n^*,$$

$$\beta | \Theta_i, d \quad \text{for } i = 1, 2, \ldots, n$$

in which n^* is the number of distinct clusters among all Θ_i, for $i = 1, 2, \ldots, n$, where Θ_j^* is the notation of the jth cluster for $j = 1, 2, \ldots, n^*$. For more details, consider a sample with a size of n. If among these n units, n_1 units are equal to a_1, n_2 units are equal to a_2, and finally, n_{n^*} units are equal to a_{n^*}, then it can be said these n units constitute n^* clusters where $\sum_{i=1}^{n^*} n_i = n$.

The vector of $c = (c_1, c_2, \ldots, c_n)$ is called the indicator vector of clusters. That is, if $c_i = j$, then the ith observation belongs to the jth cluster (also it is notated with $\Theta_i = \Theta_j^*$).

Our goal in this section is to use the Gibbs sampling method to fit Model 3.5. After calculating the conditional posterior distribution of the above conditional variables, the simulation of the parameters with any desired volume will be feasible using the following five stages. In the next chapter, we will explain all the stages by considering a particular kernel.

Stage one: Simulating Θ_i from the following posterior density function.

$$f\left(\Theta_i | \{(\Theta_{i'}, c_{i'}), i \neq i'\}, \xi, \beta, d\right), \quad i = 1, 2, \ldots, n \qquad (4.24)$$

Stage two: Determining the cluster of Θ_j^* by sampling from the following posterior density function.

$$f\left(\Theta_j^* | c, \xi, \beta, d\right), \quad j = 1, 2, \ldots, n^* \qquad (4.25)$$

Stage three: Updating all hyperparameters, $(\xi_i)_{i \geq 0}$, and the precision parameter α with the use of their relevant posterior density functions.

Stage four: Generating new samples of β from its relevant posterior density function.

Stage five: Computation of $F(t|\beta, G, x_0)$ that is the CDF of the failure time at the normal stress level x_0.

It should be noted that the first stage is performed completely independent of other stages. However, stages two, three, and four are repeated within a loop until the desired volume is reached. The generation of each of these parameters depends on the previous values of the other parameters. Stage five is calculated after completing the preceding four stages. Pay attention to the following stages.

4.6.1 First Stage: Updating Θ_i

In the first step, we first generate an initial value for the precision parameter and the hyperparameters in the model from their prior distribution. Then, according to these initial values, we extract the new values of the parameter Θ_i, for $i = 1, 2, \ldots, n$, from the prior distribution according to the base distribution and the Dirichlet process. Then, the indicator variable c_i can be updated with respect to the ith observation (t_i). According to the Blackwell–MacQueen model, a newly generated unit of Θ_i may be repetitive from the previous sample (i.e., $\Theta_{i'}$, $i' \neq i$) or new value may come from the base distribution of the Dirichlet process.

Suppose n^{*-} is the number of clusters that have been made by $\Theta_{i'}$, $i' \neq i$. More precisely, n^{*-} is the number of clusters created by the following vector:

$$(\Theta_1, \ldots, \Theta_{i-1}, \Theta_{i+1}, \ldots, \Theta_n)$$

In addition, suppose Θ_j^{*-}, for $j = 1, 2, \ldots, n^{*-}$, comes from n^{*-} clusters that are created by different values of $\Theta_{i'}$, $i' \neq i$. Note that n_j^- is the number of sample units into the n_j^{*-}th cluster for $j = 1, 2, \ldots, n^{*-}$.

If we suppose t_i is a complete failure time, namely $\delta_i = 1$, then according to equation (4.19) the conditional posterior distribution of parameter Θ_i is as follows:

$$f\left(\Theta_i | \{(\Theta_{i'}, c_{i'}), i \neq i'\}, \xi, \beta, t_i, x_i\right) = \frac{q_0^O h^O(\Theta_i | \xi, \beta, t_i, x_i) + \sum_{j=1}^{n^{*-}} n_j^- q_j^O \Delta_{\Theta_j^{*-}}}{q_0^O + \sum_{j=1}^{n^{*-}} n_j^- q_j^O}$$

(4.26)

where $\Delta_{\Theta_j^{*-}}$ is the point mass in the point Θ_j^{*-}, and both q_j^O and q_0^O are calculated as follows:

$$q_j^O = f\left(t_i | \Theta_j^{*-}\right),$$

$$q_0^O = \alpha \int_\Theta f(t_i | \Theta_i) G_0(\Theta_i) d\Theta_i$$

(4.27)

It is worth noting that the superscript "O" refers to the complete failure time and "c" refers to censored data. According to equation (4.26), Θ_i is equal to Θ_j^{*-} (a repetitive value) with the probability of

$$\frac{n_j^- q_j^O}{\left(q_0^O + \sum_{j=1}^{n^{*-}} n_j^- q_j^O\right)}$$

(4.28)

And is a new vale that is generated from $h^O(\Theta_i | \xi, \beta, t_i, x_i)$ with the following proportion:

$$\frac{q_0^O}{\left(q_0^O + \sum_{j=1}^{n^{*-}} n_j^- q_j^O\right)}$$

(4.29)

where

$$h^O(\Theta_i | \xi, \beta, t_i, x_i) \propto f(t_i | \Theta_i) G_0(\Theta_i)$$

(4.30)

If we suppose t_i is a right-censored failure-time, namely $\delta_i = 0$, then the conditional posterior distribution of parameter Θ_i can be written as follows. Note that in this case, the reliability function of that kernel distribution will be used, instead of the kernel of density. Therefore, the posterior density function of the parameters can be achieved as follows when the censored data are examined:

$$f\left(\Theta_i | \{(\Theta_{i'}, c_{i'}), i \neq i'\}, \xi, \beta, t_i, x_i\right) = \frac{q_0^c h^c(\Theta_i | \xi, \beta, t_i, x_i) + \sum_{j=1}^{n^{*-}} n_j^- q_j^c \Delta_{\Theta_j^{*-}}}{q_0^c + \sum_{j=1}^{n^{*-}} n_j^- q_j^c}$$

(4.31)

where q_j^c and q_0^c are defined as follows:

$$q_j^c = 1 - F\left(t_i | \Theta_j^{*-}\right),$$

$$q_j^c = \alpha \int_\Theta \left[1 - F(t_i|\Theta_i)\right] G_0(\Theta_i) d\Theta_i \qquad (4.32)$$

and

$$h^c(\Theta_i | \xi, \beta, t_i, x_i) \propto \left[1 - F(t_i|\Theta_i)\right] G_0(\Theta_i) \qquad (4.33)$$

After simulation of parameter Θ_i, for $i = 1,2,\ldots, n$, by applying complete or censored dataset, the cluster of the ith generated data (i.e., c_i) must be specified prior to run the next stage. This can be done using the Chinese restaurant process. The codes of this simulation have been completed by using the R programming language. This can be found in the Appendix B.9.

4.6.2 Second Stage: Updating Θ_j^*

After completing the first stage for n observation, we need to determine and update the cluster vector, c, as well as the cluster position namely Θ_j^* for $j = 1,2,\ldots, n^*$. Note that the goal of running this stage is to determine the cluster positions, Θ_j^*, which will occur if and only if ξ, c, d are known.

Suppose t_j^O, t_j^c are, respectively, the set of complete and censored data relating to the jth cluster and are defined as follows:

$$\begin{aligned} t_j^O &= \{t_i, c_i = j, \delta_i = 1\} \\ t_j^c &= \{t_i, c_i = j, \delta_i = 0\} \end{aligned} \qquad (4.34)$$

Besides, the total dataset (both complete and censored) in the jth cluster can be shown as follows:

$$t_j = t_j^O \cup t_j^c \equiv \{t_i, c_i = j\} \qquad (4.35)$$

The conditional posterior distribution of Θ_j^* is generally formulated using Bayes' law as follows:

$$f\left(\Theta_j^* | \xi, \beta, c, d\right) \propto G_0\left(\Theta_j^* | \xi\right) \times \prod_{t_i \in t_j^O} f\left(t_i | \Theta_j^*\right) \times \prod_{t_i \in t_j^c} \left[1 - F\left(t_i | \Theta_j^*\right)\right] \qquad (4.36)$$

Contrary to the simplicity of some problems, it is sometimes necessary to adopt specific methods to find the conditional posterior distribution of each parameter to generate samples from them. Naturally, each of them has its problems. In the next chapter, we will clarify how to implement all the simulation stages listed in this chapter by applying a specific example.

4.6.3 THIRD STAGE: UPDATING ξ

To update all hyperparameters and also the precision parameter, it is also necessary to calculate their posterior distributions. Since the calculation of the posterior distribution requires the prior distribution, we examine this stage in a practical and accurate way using an example in the next chapter.

4.6.4 FOURTH STAGE: UPDATING β

Updating the parameter β, like other parameters, is done using the conditional posterior distribution. The conditional posterior distribution of this parameter for complete and incomplete (censored) data can be calculated as follows:

$$f\left(\beta|\{\Theta_i, i=1,2,\ldots,n\}, d\right) \propto$$
$$f(\beta|\xi_i, i \geq 0) \times \prod_{\{i,\, \delta_i=1\}} f(t_i|\Theta_i) \times \prod_{\{i,\, \delta_i=0\}} \left[1 - F(t_i|\Theta_i)\right] \quad (4.37)$$

Now if there is a kernel function, as well as the prior density of the parameter β, the conditional posterior distribution can be obtained and, finally, the β values can be updated.

4.6.5 FIFTH STAGE: UPDATING THE DISTRIBUTION OF FAILURE-TIME

One of the goals of the ALT methods is to predict the distribution of failure time. Because researchers are looking for a way to improve estimation methods, there are many developed ways to estimate this key function.

Kuo and Mallick (1997) have considered two methods for predicting the CDF of failure time at the normal stress level x_0. In the first method, a sample of the failure time at the normal stress level of x_0 is simulated. The CDF of the failure time is then predicted using simulated data and some empirical methods such as the Kaplan–Meier estimator. The second method is performed using the kernel function form. This means that by combining the samples obtained from the multiple chain method in the Gibbs sampling, the cumulative distribution function is obtained.

The method we use here is based on the method used by Kottas (2006).

In this method, the first to fourth steps are fully implemented. In the fifth step, a value for the CDF at the normal stress level x_0 is simulated. To be more precise, consider the equation of the distribution function of the failure time. This equation can be rewritten at the normal stress level as follows:

$$F(t \mid \beta, G, x_0) = \int F(t \mid \Theta, x_0, \beta) G(d\,\Theta) \quad (4.38)$$

Suppose L is the number of components in the mixture model. Now, using this rewritten equation, in any iteration of the simulation algorithm, a new value for CDF would be predicted at the normal stress level x_0. According to Sethuraman (1994), using the Dirichlet process mixture model, the above CDF can be estimated using the finite mixture model by considering a large number of iterations in the creation of the new sample as follows:

$$F(t_m \mid \beta, G, x_0) \approx \sum_{l=1}^{L} \varpi_l F(t_m \mid \Theta_{l'}, x_0, \beta) \qquad (4.39)$$

where t_m, for $m = 1, 2, \ldots, N$, is an observation that comes from the domain of $F(t \mid \beta, G, x_0)$.

With respect to the Blackwell–MacQueen urn scheme explained in Chapter 3, $\Theta_{l'}$ for $l' = 1, 2, \ldots, L$ are independent and are getting generated from the following function:

$$\frac{\alpha}{\alpha + n} G_0(\Theta \mid \xi) + \frac{1}{\alpha + n} \sum_{i=1}^{n} \Delta_{\{\Theta_i\}} \qquad (4.40)$$

We use the method developed by Sethuraman (1994) to estimate the weight coefficients ϖ_l for $l = 1, 2, \ldots, L$. To use this method, first of all, there is a need to simulate variable z_l from $\text{Beta}(1, \alpha + n)$. Then, ϖ_l can be simulated by applying the following algorithm:

$$\varpi_1 = z_1,$$

$$\varpi_l = z_l \prod_{s=1}^{L-1} z_s; \quad l = 1, 2, \ldots, L-1, \qquad (4.41)$$

$$\varpi_L = 1 - \sum_{j=1}^{L-1} \varpi_j$$

This method is based on the definition of the Dirichlet process and the structure of this process, which Sethuraman (1994) has also used.

In summary, in Chapter 4, we first introduced the mixture models and then tried to introduce a regression model called semiparametric log-linear regression, which is widely used in ALTs. In the next section, we briefly studied the methods of determining the base distribution as well as the precision parameter of the Dirichlet process. We then explained the Dirichlet process mixture model with a kernel. Finally, we tried to clarify the simulation algorithm for estimating unknown parameters using the MCMC method.

In Chapter 5, we will use the Weibull kernel to demonstrate the goodness of the performance of the Dirichlet process mixture model in the ALTs. The results of this specific model will be illustrated by applying two different examples. In order to observe the appropriateness of the estimated model, it is necessary to compare this model with the empirical estimation distribution function of the data, which is obtained nonparametrically and without any restrictive assumption. In addition, a parametric model whose parameters are estimated by the maximum likelihood method has been adopted to compare with this nonparametric Bayesian model.

5 Illustrative Examples and Results

In this chapter, we will use the Weibull kernel to examine the performance of the Dirichlet process mixture model by applying two illustrative examples with real data. To examine the appropriateness of the estimated model, it is necessary to compare this model with the empirical estimation of data distribution which has been obtained nonparametrically and without any restrictive assumption. Besides, a parametric model for the accelerated lifetime test (ALT), in which the parameters have been estimated by the maximum likelihood method, has also been used to compare with the original model proposed in this chapter.

Since the empirical distribution function is often considered as the best criterion to assess the goodness of a selected model, we first briefly review the empirical distribution function in this chapter. Then, we try to rewrite the general model (equation 3.5), which is referred to in Chapter 4, with respect to the Weibull kernel.

We then turn the general algorithm of the previous chapter into a special case, and finally, using two real examples, we attempt to present a comprehensive analysis to compare the main model of the present book, namely the Dirichlet process mixture model which is a nonparametric model, with a parametric model.

5.1 EMPIRICAL DISTRIBUTION FUNCTION

The empirical distribution function is an estimate of the distribution function and is widely used in nonparametric studies. Generally, empirical distribution (also named empirical distribution function) can be used to describe and study a specific variable by applying a relevant sample of observations. Its value at a given point is equal to the ratio of the number of observations of the sample that are less than or equal to the given point divided by the sample size.

Suppose n products enter an experiment, and after the failure of the last unit, their lifetime is recorded like $t_1, t_2, \ldots, t_n$. If we consider $F(t)$ as the unknown cumulative distribution function of these n observations, then the nonparametric estimation of this function, also showed with $\hat{F}(t)$, can be calculated using the following simple formula:

$$\hat{F}(t_{(i)}) = \frac{i}{n}, \quad i = 1, 2, \ldots, n \tag{5.1}$$

where $t_{(i)}$ is the ith ordered observation, that is, $t_{(i)}$ is the ith lowest observation among the n data. It has been proved in probability theory that $\hat{F}_n(t)$ tends to $F(t)$ when $n \to \infty$.

Therefore, for samples big enough, it can be written that

$$F(t_{(i)}) \cong \hat{F}_n(t) \cong \frac{i}{n}, \quad i = 1, 2, \ldots, n \tag{5.2}$$

When it comes to drawing probability charts, since $F^{-1}\left(\frac{n}{n}\right) = \infty$, we will face trouble. To avoid this problem, $\frac{i}{n+1}$ is often used instead of $\frac{i}{n}$. In many reliability studies, the fraction $\frac{i-0.5}{n}$ is often considered as the nonparametric estimator of the cumulative distribution function. The fraction $\frac{i-0.3}{n+0.4}$ has been applied in this book.

5.2 DIRICHLET PROCESS WEIBULL MIXTURE MODEL

According to the following log-linear regression model,

$$T_i = e^{-x_i'\beta + W_i} = e^{-x_i'\beta} V_i$$

Kuo and Mallick (1997) carried out the Bayesian analysis of ALTs using the Dirichlet process mixture model by applying the normal and the log-normal distribution kernels. Kottas (2006) expanded their research by applying a Weibull kernel.

Ghosh et al. (2007) tried to focus on Kottas's work. Hence, they continued Kottas's research by applying the Weibull kernel but with a constant shape parameter. The method addressed in the present book is based on researches released by Kuo and Mallick (1997), Kottas (2006), and Ghosh et al. (2007). In general, we try to present a step-by-step solution to the nonparametric Bayesian analysis of accelerated lifetime data using the Dirichlet process Weibull mixture model.

5.2.1 DETERMINING BASE DISTRIBUTION

To present a clear path of simulation using the Dirichlet process mixture model, it is necessary to determine a known base distribution. The base distribution used in this book is in accordance with the base distribution used by Kottas, which is defined as follows:

$$G_0(\alpha, \lambda | \phi, \gamma) = \text{Uniform}(\alpha | 0, \phi) \times \text{Inverse} - \text{Gamma}(\lambda | d, \gamma) \tag{5.3}$$

In this chapter, the precision parameter of the Dirichlet process will be shown with μ. In addition, the parameters of the kernel distribution, namely Weibull, are notated with (α, λ). Therefore, the vector of parameters $\Theta = (\theta_1, \theta_2, \ldots, \theta_k)$ diminishes to $\Theta = (\alpha, \lambda)$.

This base distribution is highly efficient in computation (because it is conjugate), and, because of the existence of the shape parameter, it is flexible.

Illustrative Examples and Results

As is clear, when the shape parameter, α, is constant, the distribution of the inverse gamma is a conditional conjugate prior for the scale parameter, λ, of the Weibull distribution. There is also no conjugate prior for the shape parameter of the Weibull distribution. To prevent restrictions on the scale parameter, we set $d = 2$ to make the variance of the inverse-gamma distribution infinite. Doing so also reflects the lack of prior knowledge of the scale parameter.

By selecting the base distribution of the dielectric process, new parameters also appear in this distribution (which are called hyperparameters). Prior distributions are necessary to be selected for these parameters, too. The prior distributions for ϕ, γ, μ, and β are as follows:

$$\phi \sim \text{Pareto}(\phi|a_\phi, b_\phi),$$

$$\gamma \sim \text{Gamma}(\gamma|a_\gamma, b_\gamma),$$

$$\mu \sim \text{Gamma}(\mu|a_\mu, b_\mu),$$

$$\beta \sim \text{Normal}(\beta|a_\beta, b_\beta)$$

(5.4)

The Pareto and gamma distributions are conjugate for ϕ and γ. Also, the prior gamma distribution for the precision parameter, μ, due to the positive range of this distribution, can have useful results in addition to simplicity in calculations (Escobar and West, 1995). The choice of the normal distribution as the prior distribution for the parameter β is due to the fact that this parameter may have negative values in addition to positive values.

With all the definitions and assumptions considered so far, in general, the hierarchical form of the model of the Dirichlet process mixture model with the Weibull kernel can be written as follows:

$$t_i \mid \alpha_i, \lambda_i, \beta \sim \begin{cases} e^{x_i\beta} k\left(t_i e^{x_i\beta}|\alpha_i, \lambda_i\right), & \text{if } \delta_i = 1, \quad i = 1, 2, \ldots, n \\ 1 - K\left(t_i e^{x_i\beta}|\alpha_i, \lambda_i\right), & \text{if } \delta_i = 1 \end{cases}$$

$$(\alpha_i, \lambda_i) \mid G \sim G(\alpha, \lambda), \quad i = 1, 2, \ldots, n$$

$$G(\alpha, \lambda) \sim \text{DP}(\mu, G_0(\alpha, \lambda)),$$

(5.5)

$$G_0(\alpha, \lambda|\phi, \gamma) = \text{Uniform}(\alpha|0, \phi) \times \text{Inverse-Gamma}(\lambda|d, \gamma),$$

$$\mu \sim \text{Gamma}(\mu|a_\mu, b_\mu)$$

Note that in this new hierarchical model, the vector of hyperparameters, $\xi = (\alpha, \xi_i)_{i \geq 1}$, converts to $\xi = (\mu, \phi, \gamma, a_\mu, b_\mu, a_\phi, b_\phi, a_\gamma, b_\gamma, a_\beta, b_\beta)$.

5.3 ASSESSING THE MODEL AND SIMULATION

In the present Dirichlet process Weibull mixture model for the accelerated lifetime data, by considering $\theta_i = (\alpha_i, \lambda_i)$, for $i = 1, 2, \ldots, n$, the conditional posterior distributions of the following conditional parameters are required to simulate vector of parameters $(\theta_1, \ldots, \theta_n, \phi, \gamma, \mu, \beta \mid d)$:

$$
\begin{aligned}
&\theta_i \mid (\theta_{-i}, c_{-i}), \phi, \gamma, \mu, \beta, d \quad \text{for } i = 1, 2, \ldots, n, \\
&\theta_j^* \mid c, n^*, \phi, \gamma, \mu, \beta, d \quad \text{for } j = 1, 2, \ldots, n^*, \\
&\mu \mid \{\theta_j^*, j = 1, 2, \ldots, n^*\}, n^*, d, \\
&\gamma \mid \{\theta_j^*, j = 1, 2, \ldots, n^*\}, n^*, d, \\
&\phi \mid \{\theta_j^*, j = 1, 2, \ldots, n^*\}, n^*, d, \\
&\beta \mid \theta_i, d \quad \text{for } i = 1, 2, \ldots, n
\end{aligned}
\tag{5.6}
$$

To obtain the conditional posterior distribution functions of the above parameters and finally to simulate samples by applying the Gibbs sampling method, it is sufficient to rewrite the five steps of the simulation algorithm referred to in the previous chapter in a special way (namely using the Weibull kernel). These steps are rewritten as follows. Note that the descriptions and details of the simulation algorithm steps are outlined in the previous chapter. Therefore, in this section, only the changes that need to be considered in a particular case (Weibull kernel) will be considered.

5.3.1 Updating (α_i, λ_i)

The changes in this stage are as follows (for complete data):

$$
\begin{aligned}
&f\left(\alpha_i, \lambda_i \mid \{(\alpha_{i'}, \lambda_{i'}, c_{i'}) i \neq i'\}, \phi, \gamma, \mu, \beta, t_i, x_i\right) \\
&= \frac{q_0^o h^o(\alpha_i, \lambda_i \mid \phi, \gamma, \mu, \beta, t_i, x_i) + \sum_{j=1}^{n^{*-}} n_j^- q_j^o \Delta_{(\alpha_j^{*-}, \lambda_j^{*-})}}{q_0^o + \sum_{j=1}^{n^{*-}} n_j^- q_j^o}
\end{aligned}
\tag{5.7}
$$

where

$$
q_j^o = k\left(t_i e^{x_i \beta} \mid \alpha_j^{*-}, \lambda_j^{*-}\right)
\tag{5.8}
$$

and

Illustrative Examples and Results

$$q_0^O = \mu \int_0^\phi \int_0^\infty k\left(t_i e^{x_i\beta}|\alpha_i, \lambda_i\right) G_0(\alpha_i, \lambda_i) d\alpha_i d\lambda_i$$

$$= \mu \int_0^\phi \int_0^\infty k\left(t_i e^{x_i\beta}|\alpha_i, \lambda_i\right) \frac{1}{\phi} \frac{\gamma^d}{\Gamma(d)} \lambda_i^{-d-1} e^{-\left(\frac{\gamma}{\lambda_i}\right)} d\alpha_i d\lambda_i \quad (5.9)$$

$$= \frac{d\mu\gamma^d}{\phi} \int_0^\phi \frac{\alpha_i \left(t_i e^{x_i\beta}\right)^{\alpha_i - 1}}{\left(\gamma + \left(t_i e^{x_i\beta}\right)^{\alpha_i}\right)^{(d+1)}} d\alpha_i$$

and

$$h^O(\alpha_i, \lambda_i|\phi, \gamma, \mu, \beta, t_i, x_i) \propto k\left(t_i e^{x_i\beta}|\alpha_i, \lambda_i\right) G_0(\alpha_i, \lambda_i)$$

$$\propto k\left(t_i e^{x_i\beta}|\alpha_i, \lambda_i\right) \frac{1}{\phi} \frac{\gamma^d}{\Gamma(d)} \lambda_i^{-d-1} e^{-\left(\frac{\gamma}{\lambda_i}\right)} I_{\{\alpha_i \in (0,\phi)\}} \quad (5.10)$$

Equation (5.10) can be written using conditional probability as follows:

$$h^O(\alpha_i, \lambda_i|\phi, \gamma, \mu, \beta, t_i, x_i) = f(\alpha_i|\phi, \gamma, \mu, \beta, t_i, x_i) f(\lambda_i|\alpha_i, \phi, \gamma, \mu, \beta, t_i, x_i) \quad (5.11)$$

In order to generate new values of (α_i, λ_i) from the posterior distribution $h^O(.)$, first, a new value of α_i must be extracted from $f(\alpha_i|\phi, \gamma, \mu, \beta, t_i, x_i)$. Then by considering the new value of α_i, the value of λ_i can be simulated from its posterior distribution namely $f(\lambda_i|\alpha_i, \phi, \gamma, \mu, \beta, t_i, x_i)$.

Now, by considering the based distribution and the prior distribution of hyperparameters, the following conditional posterior distribution should be used to generate (α_i, λ_i):

$$f(\alpha_i|\phi, \gamma, \mu, \beta, t_i, x_i) = \int_0^\infty h^O(\alpha_i, \lambda_i|\phi, \gamma, \mu, \beta, t_i, x_i) d\lambda_i$$

$$\propto \frac{\alpha_i \left(t_i e^{x_i\beta}\right)^{\alpha_i - 1}}{\left(\gamma + \left(t_i e^{x_i\beta}\right)^{\alpha_i}\right)^{(d+1)}} I_{\{\alpha_i \in (0,\phi)\}} \quad (5.12)$$

and

$$f(\lambda_i|\alpha_i, \phi, \gamma, \mu, \beta, t_i, x_i) \propto \lambda_i^{-d-2} e^{-\left(\frac{\gamma + \left(t_i e^{x_i\beta}\right)^{\alpha_i}}{\lambda_i}\right)} \quad (5.13)$$

$$\approx \text{Inverse} - \text{Gamma}\left(d+1, \gamma + \left(t_i e^{x_i\beta}\right)^{\alpha_i}\right)$$

Generation of λ_i can be done using the inverse-gamma distribution. As regards α_i, that there is no closed form of its density function, there are several methods to generate this parameter.

For example, the accept-reject and the distribution function methods are useful ways to extract α_i from its density function. The method used in this book to generate α_i from the determined density function is the slice sampling method (Walker, 2006; Neal, 2003, 2008).

These conditional posterior density functions can be written for censored data as follows.

Suppose t_i, for $i = 1, 2, \ldots, n$, is a sample of right-censored data. Therefore, in this case, $\delta_i = 0$ and the Weibull reliability function should replace the Weibull density function. The conditional posterior density function of (α_i, λ_i) can be achieved as follows:

$$f\left(\alpha_i, \lambda_i \big| \{(\alpha_{i'}, \lambda_{i'}, c_{i'}) i \neq i'\}, \phi, \gamma, \mu, \beta, t_i, x_i\right)$$

$$= \frac{q_0^c h^c(\alpha_i, \lambda_i | \phi, \gamma, \mu, \beta, t_i, x_i) + \sum_{j=1}^{n^{*-}} n_j^- q_j^c \Delta_{(\alpha_j^{*-}, \lambda_j^{*-})}}{q_0^c + \sum_{j=1}^{n^{*-}} n_j^- q_j^c} \quad (5.14)$$

where

$$q_j^c = 1 - K\left(t_i e^{x_i \beta} | \alpha_j^{*-}, \lambda_j^{*-}\right),$$

$$q_0^c = \mu \int_0^\phi \int_0^\infty \left[1 - K\left(t_i e^{x_i \beta} | \alpha_i, \lambda_i\right)\right] G_0(\alpha_i, \lambda_i) \, d\alpha_i d\lambda_i$$

$$= \mu \int_0^\phi \int_0^\infty \left[1 - K\left(t_i e^{x_i \beta} | \alpha_i, \lambda_i\right)\right] \frac{1}{\phi} \frac{\gamma^d}{\Gamma(d)} \lambda_i^{-d-1} e^{-\left(\frac{\gamma}{\lambda_i}\right)} d\alpha_i d\lambda_i \quad (5.15)$$

$$= \frac{\mu \gamma^d}{\phi} \int_0^\phi \frac{1}{\left(\gamma + \left(t_i e^{x_i \beta}\right)^{\alpha_i}\right)^{(d+1)}} d\alpha_i$$

and

$$h^c(\alpha_i, \lambda_i | \phi, \gamma, \mu, \beta, t_i, x_i)$$

$$\propto \left[1 - K\left(t_i e^{x_i \beta} | \alpha_i, \lambda_i\right)\right] G_0(\alpha_i, \lambda_i) \quad (5.16)$$

$$\propto \left[1 - K\left(t_i e^{x_i \beta} | \alpha_i, \lambda_i\right)\right] \frac{1}{\phi} \frac{\gamma^d}{\Gamma(d)} \lambda_i^{-d-1} e^{-\left(\frac{\gamma}{\lambda_i}\right)} I_{\{\alpha_i \in (0, \phi)\}}$$

Likewise, in order to generate the new values of (α_i, λ_i) from the posterior distribution $h^c(.)$, first, a new value of α_i must be extracted from $f(\alpha_i | \phi, \gamma, \mu, \beta, t_i, x_i)$.

Illustrative Examples and Results

Then by considering the new value of α_i, the value of λ_i can be simulated from its posterior distribution namely $f(\lambda_i|\alpha_i, \phi, \gamma, \mu, \beta, t_i, x_i)$.

In the case of censored data, the conditional posterior density function of α_i and λ_i are as follows:

$$f(\alpha_i|\phi,\gamma,\mu,\beta,t_i,x_i) = \int_0^\infty h^c(\alpha_i,\lambda_i|\phi,\gamma,\mu,\beta,t_i,x_i)\,d\lambda_i$$

$$\propto \frac{1}{\left(\gamma+\left(t_i e^{x_i\beta}\right)^{\alpha_i}\right)^d} I_{\{\alpha_i \in (0,\phi)\}} \tag{5.17}$$

and

$$f(\lambda_i|\alpha_i,\phi,\gamma,\mu,\beta,t_i,x_i) \propto \lambda_i^{-d-1} e^{-\left(\frac{\gamma+\left(t_i e^{x_i\beta}\right)^{\alpha_i}}{\lambda_i}\right)} \tag{5.18}$$

$$\approx \text{Inverse-Gamma}\left(d, \gamma+\left(t_i e^{x_i\beta}\right)^{\alpha_i}\right)$$

After applying this algorithm for both complete and right-censored data, and after updating (α_i, λ_i), it is time to determine the cluster of ith simulated parameter, c_i. This can be done using the Chinese restaurant process. The programming codes of this stage can be found in the Appendix B.9.

5.3.2 Updating $\left(\alpha_j^*, \lambda_j^*\right)$

The general equation of posterior probability density function of $\left(\alpha_j^*, \lambda_j^*\right)$ is as follows:

$$f\left(\alpha_j^*,\lambda_j^*|\phi,\gamma,\beta,c,d\right) \propto G_0\left(\alpha_j^*,\lambda_j^*|\phi,\gamma\right)$$

$$\times \prod_{t_i \in t_j^O} k\left(t_i e^{x_i\beta}|\alpha_j^*,\lambda_j^*\right) \times \prod_{t_i \in t_j^c}\left[1-K\left(t_i e^{x_i\beta}|\alpha_j^*,\lambda_j^*\right)\right] \tag{5.19}$$

In order to generate the new values of $\left(\alpha_j^*, \lambda_j^*\right)$ from posterior density $f\left(\alpha_j^*, \lambda_j^*|\phi, \gamma, \beta, c, d\right)$, first, a new value of λ_j^* must be extracted from its conditional posterior density function. Then by considering the new value of λ_j^*, the value of α_j^* can be simulated from its posterior distribution.

The conditional posterior density function of α_j^* and λ_j^* are as follows:

$$f\left(\lambda_j^*|\alpha_j^*,\phi,\gamma,\mu,\beta,c,d\right) \propto \left(\lambda_j^*\right)^{-d-1} e^{-\left(\frac{\gamma}{\lambda_j^*}\right)} \times \prod_{t_i \in t_j^O} \lambda_j^{*-1} e^{\left[\lambda_j^{*-1}\left(t_i e^{x_i\beta}\right)^{\alpha_j^*}\right]} \times \prod_{t_i \in t_j^c} e^{\left[-\lambda_j^{*-1}\left(t_i e^{x_i\beta}\right)^{\alpha_j^*}\right]}$$

$$\propto \lambda_j^{*\left(-d-n_j^O-1\right)} e^{\left[-\lambda_j^{*-1}\left(\gamma+\sum_{t_i \in t_j}\left(t_i e^{x_i\beta}\right)^{\alpha_j^*}\right)\right]}$$

$$\approx \text{Inverse} - \text{Gamma}\left(d + n_j^o, \gamma + \sum_{t_i \in t_j} \left(t_i e^{x_i \beta}\right)^{\alpha_j^*}\right) \quad (5.20)$$

and

$$f\left(\alpha_j^* | \lambda_j^*, \phi, \gamma, \mu, \beta, c, d\right) \propto \left(\alpha_j^*\right)^{n_j^o} \times I_{\{\alpha_j^* \in (0,\phi)\}} \times \prod_{t_i \in t_j^o} \left(t_i e^{x_i \beta}\right)^{\alpha_j^*} \times \prod_{t_i \in t_j} e^{\left[-\lambda_j^{*-1}\left(t_i e^{x_i \beta}\right)^{\alpha_j^*}\right]} \quad (5.21)$$

where n_j^o is the number of complete units into the jth cluster.

Because of the complex and non-closed density function of α_j^*, it is hard to generate data from this function. The method utilized here is based on the idea developed by Damien et al. (1999). This method, which also contains Markov chain properties, is treated as one of the Markov chain Monte Carlo (MCMC) methods. In addition, the application of this method is when the probability density function of the considered variable is complex and nonstandard.

To use this method, first, variables U_1 and U_2 should be defined as follows:

$$U_1 \equiv \{u_{1,i} : t_i \in t_j^o\}$$
$$U_2 \equiv \{u_{2,i} : t_i \in t_j\} \quad (5.22)$$

where U_1 is the set of complete data in jth cluster and U_2 refers to all data (whether complete or censored) belonging to jth cluster. The joint density function of (α_j^*, U_1, U_2) can be obtained as follows:

$$f\left(\alpha_j^*, u_1, u_2 | \lambda_j^*, \phi, \gamma, \mu, \beta, c, d\right) \propto \left(\alpha_j^*\right)^{n_j^o} \times I_{\{\alpha_j^* \in (0,\phi)\}}$$
$$\times \prod_{t_i \in t_j^o} I_{\left\{0 < u_{1,i} < \left(t_i e^{x_i \beta}\right)^{\alpha_j^*}\right\}} \times \prod_{t_i \in t_j} I_{\left\{0 < u_{2,i} < e^{\left[-\lambda_j^{*-1}\left(t_i e^{x_i \beta}\right)^{\alpha_j^*}\right]}\right\}} \quad (5.23)$$

According to joint density function (4.4), $u_{1,i}$ and $u_{2,i}$ first should be generated from the following density functions:

$$U_{1,i} \sim \text{Uniform}\left(0, \left(t_i e^{x_i \beta}\right)^{\alpha_j^*}\right)$$
$$U_{2,i} \sim \text{Uniform}\left(0, e^{\left[-\lambda_j^{*-1}\left(t_i e^{x_i \beta}\right)^{\alpha_j^*}\right]}\right) \quad (5.24)$$

Illustrative Examples and Results

Then α_j^* should be generated from the conditional posterior density function as follows:

$$f\left(\alpha_j^*|u_1,u_2,\lambda_j^*,\phi,\gamma,\mu,\beta,c,d\right) \propto \left(\alpha_j^*\right)^{n_j^O} \tag{5.25}$$

According to equation (5.23), there are three domains for simulating $\left(\alpha_j^*, U_1, U_2\right)$ as follows:

$$\begin{aligned}\alpha_j^* &\in (0,\phi) \\ u_{1,i} &\in \left(0,\left(t_i e^{x_i\beta}\right)^{\alpha_j^*}\right) \\ u_{2,i} &\in \left(0, e^{\left[-\lambda_j^{*-1}\left(t_i e^{x_i\beta}\right)^{\alpha_j^*}\right]}\right)\end{aligned} \tag{5.26}$$

Hence, a new simulated α_j^* that has been extracted from conditional density function (4.5) is acceptable if and only if it applies in these three areas. More precisely, if α_j^* gets simulated from equation (4.5), it should fall in the following set:

$$(0,\phi) \cap \left(0,\left(t_i e^{x_i\beta}\right)^{\alpha_j^*}\right) \cap \left(0, e^{\left[-\lambda_j^{*-1}\left(t_i e^{x_i\beta}\right)^{\alpha_j^*}\right]}\right) \tag{5.27}$$

According to the algorithm described in the second stage, the position of the clusters can be determined and new data can be simulated to the desired number using the Gibbs sampling method.

5.3.3 Updating ϕ, γ, and μ

5.3.3.1 Updating ϕ

As mentioned in Chapter 4, the prior density function of parameter ϕ is a Pareto as follows:

$$f\left(\phi|a_\phi,b_\phi\right) = \frac{a_\phi b_\phi^{a_\phi}}{\phi^{a_\phi+1}}, \quad \phi > 0, \quad a_\phi > 0, \quad b_\phi > 0 \tag{5.28}$$

Therefore, using Bayes' law, the conditional posterior distribution of this parameter can be computed as follows:

$$f\left(\phi|\alpha_j^*, \lambda_j^*, n^*\right) \propto \prod_{j=1}^{n^*} f(\alpha_j^* \mid \phi, n^*) f(\phi|a_\phi, b_\phi)$$

$$\propto \frac{1}{\phi^{n^*}} \frac{a_\phi b_\phi^{a_\phi}}{\phi^{a_\phi+1}} \tag{5.29}$$

$$\propto \frac{1}{\phi^{a_\phi+n^*+1}} I_{\{\phi > \max\{b_\phi, \max\{\alpha_j^*, j=1,2,\ldots, n^*\}\}\}}$$

$$\approx \text{Pareto}\left(a_\phi + n^*, \max\left\{b_\phi, \max\left\{\alpha_j^*, j=1,2,\ldots, n^*\right\}\right\}\right)$$

Finally, to generate a sample of parameter ϕ, it is enough to use this Pareto density function as the conditional density function.

5.3.3.2 Updating γ

The considered method to update this parameter is similar to the approach applied for ϕ but with a different prior density function.

According to the general algorithm introduced in Chapter 4, the prior density function of γ is a gamma distribution is defined as follows:

$$f\left(\gamma|a_\gamma, b_\gamma\right) = \frac{\gamma^{a_\gamma-1} e^{-\gamma b_\gamma} b_\gamma^{a_\gamma}}{\Gamma(a_\gamma)}, \quad \gamma > 0,\ a_\gamma > 0,\ a_\gamma > 0 \tag{5.30}$$

Therefore, the conditional posterior density function of this parameter can also be achieved as follows:

$$f\left(\gamma|\alpha_j^*, \lambda_j^*, n^*\right) \propto \prod_{j=1}^{n^*} f(\lambda_j^* \mid \gamma, n^*) f(\gamma|a_\gamma, b_\gamma)$$

$$\propto \gamma^{n^* d} e^{-\gamma \sum_{j=1}^{n^*} (\lambda_j^*)^{-1}} \frac{\gamma^{a_\gamma-1} e^{-\gamma b_\gamma} b_\gamma^{a_\gamma}}{\Gamma(a_\gamma)} \tag{5.31}$$

$$\propto \gamma^{n^* d + a_\gamma - 1} e^{-\gamma\left(b_\gamma + \sum_{j=1}^{n^*} (\lambda_j^*)^{-1}\right)}$$

$$\approx \text{Gamma}\left(n^* d + a_\gamma, \left(b_\gamma + \sum_{j=1}^{n^*} (\lambda_j^*)^{-1}\right)\right)$$

Finally, to generate a sample of parameter γ, it is enough to use this gamma density function as the conditional density function.

5.3.3.3 Updating μ

To update the precision parameter using the conditional posterior density function, the method developed by Escobar and West (1995) will be applied. As mentioned, the gamma distribution is an appropriate prior density function to this parameter.

Their method is based on an auxiliary random variable. The steps to update this parameter are as follows:

$$P(n^*|n) = \int P(n^*, \mu|n) d\mu$$
$$= \int P(n^*|\mu, n) P(\mu|n) d\mu \tag{5.32}$$
$$= E_{P(\mu|n)}\left(P(n^*|\mu, n)\right), \quad n^* = 1, 2, \ldots, n$$

Now, with reference to the endeavors of Antoniak (1974), consider Π as the vector of coefficients of the mixture model. With sampling from the density function of Π, we, in fact, update the number of clusters (n^*).

Consider the following relations,

$$P(n^*|\mu, n) = c_n(n^*) n! \mu^{n^*} \frac{\Gamma(\mu)}{\Gamma(\mu+n)}, \quad n^* = 1, 2, \ldots, n \tag{5.33}$$

where

$$c_n(n^*) = P(n^*|\mu = 1, n) \tag{5.34}$$

To calculate different coefficients of $c_n(n^*)$, Stirling numbers can be used.

Now, if we suppose Π and n^* are known, then

$$P(\mu|n^*, \Pi, d) = P(\mu|n^*) \propto P(\mu) P(n^*|\mu) \tag{5.35}$$

Equation (5.33) can be rewritten as follows:

$$\frac{\Gamma(\mu)}{\Gamma(\mu+n)} = \frac{(\mu+n)\beta(\mu+1,n)}{\mu\Gamma(n)} \tag{5.36}$$

where $\beta(.,.)$ is the beta distribution.

Now, with reference to equation (5.35), for $n^* = 1, 2, \ldots, n$:

$$P(\mu|n^*) \propto P(\mu) \mu^{(n^*-1)} (\mu+n) \beta(\mu+1,n)$$
$$\propto P(\mu) \mu^{(n^*-1)} (\mu+n) \int_0^1 x^\mu (1-x)^{(n-1)} dx \tag{5.37}$$

This refers to the fact that $P(\mu|n^*)$ is the marginal probability density function of a joint distribution of μ and a variable like U. This means

$$P(\mu, u|n^*) \propto P(\mu)\mu^{(n^*-1)}(\mu+n)u^\mu(1-u)^{(n-1)}, \quad \mu > 0, \ 0 < u < 1 \quad (5.38)$$

Thus, the conditional distribution of variable $u|\mu$ is a beta distribution with parameters $(\mu+1, n)$. Now, considering the gamma distribution as the prior distribution of parameter μ, the conditional posterior distribution of this parameter can also be calculated as follows:

$$P(\mu|u, n^*, c, d) \propto \mu^{a_\mu+n^*-2}(\mu+n)e^{-\mu(b_\mu-\log(u))}$$
$$\propto \mu^{a_\mu+n^*-1}e^{-\mu(b_\mu-\log(u))} + n\mu^{a_\mu+n^*-2}e^{-\mu(b_\mu-\log(u))}, \quad \mu > 0 \quad (5.39)$$

According to this relation, obviously, the conditional posterior distribution of μ is a mixture of two different gamma distributions as follows:

$$\mu | u, n^*, c, d \sim \pi \text{Gamma}(a_\mu + n^*, b_\mu - \log(u))$$
$$+ (1-\pi)\text{Gamma}(a_\mu + n^* - 1, b_\mu - \log(u)) \quad (5.40)$$

where

$$\pi = \frac{a_\mu + n^* - 1}{n(b_\mu - \log(u)) + a_\mu + n^* - 1} \quad (5.41)$$

Therefore, to update the precision parameter of the Dirichlet process, it is enough to sample from this mixture gamma density function.

5.3.3.4 Updating β

The considered method to update this parameter is similar to the approach applied for the previous parameters but with a different prior density function. Note that the calculation of the conditional posterior density function of this parameter will be carried out by the breakdown of the complete and censored dataset.

$$f(\beta|\{\alpha_i, \lambda_i, i=1,2,\ldots,n\}, d)$$
$$\propto f(\beta|a_\beta, b_\beta) \times \prod_{\{i:\delta_i=1\}} e^{x_i\beta} k(t_i e^{x_i\beta}|\alpha_i, \lambda_i) \quad (5.42)$$
$$\times \prod_{\{i:\delta_i=0\}} \left[1 - K(t_i e^{x_i\beta}|\alpha_i, \lambda_i)\right]$$

Illustrative Examples and Results

This conditional posterior density function does not hold a closed form. Hence, the method developed by Damien et al. (1999) will be used in this case, too. Thus, two auxiliary variables are required to run this method.

If the lifetime data is complete, i.e., $i \in \{i : \delta_i = 1\}$, then the auxiliary variable $u_{1,i}$ should be generated from a uniform distribution in the following interval:

$$\left(0, \, e^{\alpha_i x_i \beta}\right) \tag{5.43}$$

And if the considered data is censored, i.e., $i \in \{i : \delta_i = 0\}$, then the auxiliary variable $u_{2,i}$ should be generated from a uniform distribution in the following interval:

$$\left(0, \, e^{-\lambda_i^{-1} t_i^{\alpha_i} e^{\alpha_i x_i \beta}}\right) \tag{5.44}$$

As mentioned in Chapter 4, the prior density function for β is a normal distribution with known parameters. To apply the method introduced by Damien et al. (1999), first, we generate a sample unit form $f(\beta \mid a_\beta, b_\beta)$, and then if the simulated value falls in the following interval, it can be considered as a simulated value of the conditional posterior density function of parameter β:

$$\left(\max_{\{i : \delta_i = 1\}} \left\{ \frac{\log(u_{1,i})}{\alpha_i x_i} \right\}, \, \min_{\{i = 1, 2, \ldots, n\}} \left\{ (\alpha_i x_i)^{-1} \log\left(-\lambda_i^{-1} t_i^{-\alpha_i} \log(u_{2,i})\right) \right\} \right) \tag{5.45}$$

In fact, this interval offers a truncated normal distribution as the conditional posterior distribution for the parameter β. Note that the truncated normal distribution is restricted in this interval.

5.4 ILLUSTRATIVE EXAMPLES

To illustrate how the Dirichlet process Weibull mixture model works when the Bayesian approach is applied, this section examines two different practical examples with real data of accelerated lifetime.

Both examples use experimental data collected at the Thin Film Nano and Microelectronics Research Laboratory at Texas A&M University, College Station. Prior density functions accompanied by their known parameter for parameters ϕ, γ, μ, and β are as follows:

$$\phi \sim \text{Pareto}(1,1),$$

$$\gamma \sim \text{Gamma}(1, 0.001),$$

$$\mu \sim \text{Gamma}(1, 0.001),$$

$$\beta \sim \text{Normal}(0, 10^6)$$

(5.46)

According to Congdon (2003), Gamma$(1, 0.001)$ and Normal$(0, 10^6)$ are the distributions that widely used in the hierarchical Bayesian analysis. Note that, according to

Yuan et al. (2014), a sensitivity analysis shows that this simulation is reliable when L exceeds 2000 (in which L refers to the number of components of the mixture model). In addition, the number of iterations is 10,000 that 5000 first simulated values have been eliminated.

The absolute error loss function has been used in this study. Hence, the median value of the conditional posterior density function will be considered as the Bayesian estimation. Also, our goal in this section is to present three different estimates of the distribution function.

The first estimate is based on the Dirichlet process mixture model with the Weibull kernel and is considered as a nonparametric model. The second model is carried out by a parametric method in which the unknown parameters are estimated by the maximum likelihood method, and finally, the distribution function is predicted at the normal stress level. The last model is the estimation of distribution function at the normal stress level using the empirical method. Note that for the goodness of fit approval, we compare each of the first and second models with the empirical model.

Example 5.4.1

This example studies the reliability of a new mixed oxides high-dielectric material for nanoelectronic applications (Luo, 2004). Metal-oxide-semiconductor capacitors made with the high-dielectric film were under electrical stresses.

Note that the observations in Table 5.1 are complete and there is no censored data. In this example, although $x_0 = 7.1\,\text{MV/cm}$ is not the normal stress level, we analyze the distribution of failure time at 7.5, 7.7, and 7.9 MV/cm levels to predict the distribution function at the stress 7.1 MV/cm. In the end, this predicted distribution function can be compared with the empirical distribution function at the level of $x_0 = 7.1\,\text{MV/cm}$.

In order to have a comparative analysis of the main model discussed in this book, we also ran a parametric ALT model. In general, according to Wu et al. (2000), it has been acknowledged that the class electromagnetic gate failure pertains to the weakest-link problem of extreme value statistics, so it seems that the Weibull distribution can be considered as a best-fitted distribution of this failure time at the given stress level.

On the other hand, according to Yuan et al. (2014) and Wu (2000), the E-model (as a function of the normal stress level and accelerated data) is a well-known lifetime-stress relationship in physics that relates the accelerated lifetime to the electrical field stress with respect to an exponential relationship given by $t_{bd} = A e^{[B(E_{bd}-1)]}$, in which t_{bd} and E are the notations for failure time and the used electrical field stress, respectively. In this equation, A and B are temperature-dependent constants, and E_{bd} is a reference electrical filed stress.

In fact, the E-model is a log-linear lifetime-stress relationship. Therefore, the Weibull distribution with the log-linear lifetime-stress relationship is a tailored manner for studying the time-dependent dielectric breakdown by applying the parametric ALT model. As a parametric ALT model, the standard MLE method is to fit the Weibull ALT log-linear model which will be used to predict the cumulative distribution function of failure time at the stress level $x_0 = \text{MV/cm}$.

TABLE 5.1
Times-to-Breakdown of MOS Capacitors Tested a Four Electrical Field Stresses

Stress (MV/cm)	Time-to-Breakdown (Seconds)								
7.9	1	2	9	12	35	46	72	74	82 107
	142	153	193	251	290	348	399	511	556 1104
	1509	1535	1756	2376	2843	3140	3514	3616	3882 4583
7.7	9	18	20	25	29	66	124	127	175 221
	249	341	362	552	630	760	782	794	906 932
	968	1378	1386	1664	1728	2229	2249	2338	4058 4986
	6312	6400	6847	8474					
7.5	39	40	77	247	253	299	311	633	666 830
	950	1060	1383	1416	1742	1843	1879	1905	2096 2337
	2532	2648	3020	3434	3947	4373	4729	5215	5614 6753
	9703	9898	10,130	110,990					
7.1	28	88	99	107	211	213	248	301	311 593
	673	702	741	911	949	1040	1439	1971	2069 2253
	2501	3547	4452	4580	4882	5657	5737	6323	7565 8209
	10,000	11,650	15,250	21,620	25,910				

Source: Authors' own table (Luo, 2004; Yuan et al., 2014).

After the parameter estimation by applying the MLE method, the estimated values of three unknown parameters α, λ, and β are 0.6605, 126.1, and 0.001569, respectively.

Figure 5.1 illustrates the predicted cumulative distribution function at the stress level of $x_0 = 7.1\,\text{MV/cm}$ using three methods including the nonparametric Bayesian approach by applying the Dirichlet process Weibull mixture modes, parametric model, and empirical prediction.

Since the cumulative distribution function estimated by the nonparametric method, as compared to the estimated CDF by the parametric method, is more similar to the empirical curve, this figure reflects the goodness of the nonparametric approach as compared to the parametric method.

Figure 5.2 shows the nonparametric probability density function of this dataset. As it shows, there are several modes in this density function. It seems what has made the nonparametric Bayesian approach more reliable than the parametric method is its capability and flexibility in predicting the multimodal functions. This capability derived from the mixture model has been applied in the nonparametric method.

Some predicted values of parameters α, λ, and β, and five predicted values of the cumulative distribution function using three mentioned models, have been presented in Tables 5.2 and 5.3.

In Table 5.3, $\widehat{F_1(t_i)}$, $\widehat{F_2(t_i)}$, and $\widehat{F_3(t_i)}$ denote cumulative distribution functions estimator using nonparametric Bayes, parametric, and empirical methods, respectively.

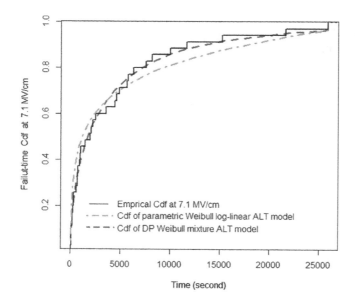

FIGURE 5.1 The predicted cumulative distribution function for accelerated lifetime data using the Dirichlet process Weibull mixture model, parametric model (herein MLE model), and the empirical method at the stress level $x_0 = 7.1$ MV/cm. (Authors' own figure (Yuan et al., 2014).)

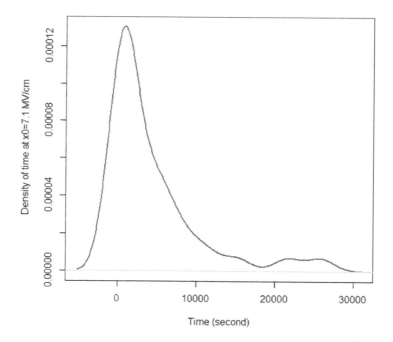

FIGURE 5.2 The nonparametric probability density function of the break-to-time data of Table 5.1 at the stress level $x_0 = 7.1$ MV/cm. (Authors' own figure.)

TABLE 5.2
Some Predicted Values of Parameters α, λ, and β with Respect to the Break-to-Time Data in Example 5.4.1

i	1	2	3	4	5
$\widehat{\alpha_i}$	0.53051	0.56516	1.32884	0.53051	0.04609
$\widehat{\lambda_i}$	70.79260	119.17715	151.50893	69.85010	140.38409
$\widehat{\beta_i}$	0.15925	0.21240	−0.09671	0.07883	0.27575

Source: Authors' own table.

TABLE 5.3
Predicted Values of Cumulative Distribution Function by the Breakdown of Three Models with Respect to the Break-to-Time Data in Example 5.4.1

i	1	2	3	4	5
$\widehat{F_1(t_i)}$	0.09073	0.16033	0.16976	0.17623	0.24037
$\widehat{F_2(t_i)}$	0.07460	0.15226	0.16352	0.17134	0.25497
$\widehat{F_3(t_i)}$	0.01977	0.04802	0.07622	0.10451	0.13276

Source: Authors' own table.

Example 5.4.2

The second example evaluates the reliability of a novel nanocrystals-embedded high-nonvolatile memory device (Lin and Kuo, 2011).

Table 5.4 is related to the failure time of memory devices in four different stress levels. In this example, the electrical voltage has been considered as the accelerator factor (also named stress factor). 7.1, 7.5, 7.9, and 8.3 V are the different stress levels.

This table includes right-censored data in four different voltage stresses. At stress levels 7.1, 7.5, and 7.9 V, there are 37, 10, and 8 right-censored data, respectively.

Like Example 5.4.1, this example tries to predict cumulative distribution function at the least stress level (namely $x_0 = 7.1\,\text{V}$) with the use of data recorded at other stress levels.

For the parameter method, the estimated values of three unknown parameters α, λ, and β are 0.04739, 0.06096, and −38.38, respectively.

Figure 5.3 illustrates the predicted curve of the cumulative distribution function of the failure time of nanocrystals with the use of nonparametric ALT method, parametric method, and empirical method. By considering the empirical cumulative distribution function as the criterion of the choice of the best model, as Figure 5.3 shows, both nonparametric ALT and parametric ALT models give the same results.

TABLE 5.4
Time-to-Break of Nanocrystals-Embedded High-k Memories Tested at Four Voltage Stresses

Stress (V)	Time-to-Breakdown (Seconds)								
8.3	2	5	5	6	6	6	10	11	15	18
	21	35	39	49						
7.9	5	8	8	12	15	26	29	39	45	69
	100	105	115	146	153	183+	(8 censored)			
7.5	5	29	31	31	33	39	46	46	54	66
	70	86	87	107	122	137	176	181	190	218
	225	259	277	334	356	371	443	480+	(10 censored)	
7.1	8	38	72	88	90	97	122	140	163	170
	188	198	199	223	232	256	257	265	318	371
	399	401	412	434	448	513	527	556	583	600+
	(37 censored)									

Source: Authors' own table (Lin and Kuo, 2011; Yuan et al., 2014).

In fact, this example shows that in some cases, the parametric method may work as good as the nonparametric Bayesian method.

To use the programming codes of Figures 5.2 and 5.3, see Section B.12 in the Appendix.

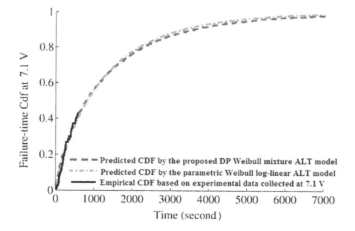

FIGURE 5.3 The predicted cumulative distribution function for accelerated lifetime data using the Dirichlet process Weibull mixture model, parametric model (herein MLE model), and the empirical method at the stress level $x_0 = 7.1$ V. (Authors' own figure (Yuan et al., 2014).)

Illustrative Examples and Results

Some estimated values of parameters α, λ, and β, and five predicted values of the cumulative distribution function using three mentioned models, have been presented in Tables 5.5 and 5.6.

In Example 5.4.1, we observed that the predicted cumulative distribution function by the nonparametric Bayesian ALT method has a higher capability than the parametric ALT method. Therefore, since the empirical distribution function is also a criterion for the goodness of a fitted model, it can be said that the distribution function which has been predicted by the nonparametric Bayes method is more accurate. Moreover, since the objective of this book is the accurate estimation of the distribution for survival data, and on the other hand, because the distribution predicted by the nonparametric Bayesian method is close to the empirical distribution, this model is expected to be useful for studying the features and properties of any dataset (especially lifetime data).

In Example 5.4.2, the results obtained from both models may be similar due to the proximity of both the nonparametric Bayesian model and the parametric method to the empirical distribution function.

What was examined in this book was a study of accelerated data with a nonparametric Bayes approach based on the Dielectric process mixture model with a desired kernel density function. Also, the goals we were looking for in Chapter 4 were achieved with the help of two real examples in Chapter 5.

TABLE 5.5

Some Predicted Values of Parameters α, λ, and β with Respect to the Break-to-Time Data in Example 5.4.2

i	1	2	3	4	5
$\widehat{\alpha_i}$	0.05487	0.06685	0.36721	1.16582	1.23783
$\widehat{\lambda_i}$	356.40700	360.60830	769.35450	1014.386460	2485.30070
$\widehat{\beta_i}$	−18.12032	−10.05552	−5.97320	−0.03704	0.073361

Source: Authors' own table.

TABLE 5.6

Predicted Values of Cumulative Distribution Function by the Breakdown of Three Models with Respect to the Break-to-Time Data in Example 5.4.2

i	1	2	3	4	5
$\widehat{F_1(t_i)}$	0.01020	0.02270	0.03917	0.05411	0.06307
$\widehat{F_2(t_i)}$	0.00981	0.02221	0.03985	0.05379	0.06891
$\widehat{F_3(t_i)}$	0.01054	0.02560	0.04066	0.05572	0.07078

Source: Authors' own table.

The MCMC algorithm was also used to estimate unknown parameters in the nonparametric model. Finally, to fit the Weibull log-linear model on complete and censored data, the R programing language was used. The slice sampling method was also used to generate data from functions that did not have a closed form. In order to increase the flexibility of the model and also to reflect the absence of prior knowledge, the prior distributions used for the hyperparameters were selected noninformative.

Despite the lack of a significant difference between the nonparametric model and the parametric model in the second example, both examples illustrated that the nonparametric Bayesian model, in addition to not imposing any restrictions on parameters, also properly estimates the distribution of products' failure time at the normal stress levels.

Although it is difficult to work with the Gibbs sampling algorithm in such models, it will be associated with considerable results. It should also be noted that after obtaining the simulation algorithm and calculations of the conditional posterior distribution functions of the parameters, sampling of the conditional posterior distribution functions using different methods will be easy if the path of simulation defines clear.

Although the results obtained in the nonparametric Bayesian analysis may be very useful, due to the difficulty of making a sampling algorithm in these models, the use of this method is not welcomed in most academic studies and researches. Besides, in nonparametric Bayesian analysis, in order to increase the accuracy of parameter estimation, there is a need for a large sample size compared to parametric analyzes. This especially will be important when the prior distribution used for the parameters is noninformative.

Notwithstanding the existence of problems and difficulties in nonparametric Bayesian analysis, it can be said that the results obtained from this method will be very satisfactory.

Appendix A
Guide to Proofs

In this section, we have tried to guide readers and prove some of the ambiguous relations used in this book.

A.1

$$\Pr(Y_1, \ldots, Y_n \mid X = j) \propto \Pr(Y_1, \ldots, Y_n) \times \Pr(X = j \mid Y_1, \ldots, Y_n)$$

$$= y_j \prod_{i=1}^{n} y_i^{a_i - 1} \tag{A.1}$$

$$= \operatorname{Dir}\left(a_1^{(j)}, a_n^{(j)}\right)$$

in which

$$a_i^{(j)} = \begin{cases} a_i & \text{if } i \neq j \\ a_i + 1 & \text{if } i = j \end{cases} \tag{A.2}$$

A.2

$$a_i^{(j)} = a_i + \sum_{i=1}^{N} \delta(x_i)$$

$$\delta(x_i) = \begin{cases} 1 & \text{if } x_i = x_j \in \chi_j \\ 0 & \text{if } x_i \neq x_j \in \chi_j \end{cases}$$

Then,

$$E(\theta_i) = \frac{a_i^{(j)}}{\sum_{i=1}^{n} a_i^{(j)}} = \frac{a_i + \sum_{i=1}^{N} \delta(x_i)}{\sum_{i=1}^{n} \left(a_i + \sum_{i=1}^{N} \delta(x_i)\right)}$$

$$= \frac{\alpha m_i + \sum_{i=1}^{N} \delta(x_i)}{\alpha + \sum_{i=1}^{n} n_i} = \frac{\alpha m_i}{\alpha + N} + \frac{1}{\alpha + N} \sum_{i=1}^{N} \delta(x_i) \tag{A.3}$$

A.3

It is enough to separate χ to two distinct partitions A and A^c. As mentioned, $G(A)$ will have a beta distribution with parameters $(\alpha G_0(A), \alpha(1-G_0(A)))$. Hence,

$$E(G(A)) = G_0(A)$$

$$\text{Var}(G(A)) = \frac{G_0(A)(1-G_0(A))}{1+\alpha}$$

A.4

With respect to the theorem assumptions, it is clear that

$$\Pr(X \in A | G(A)) = G(A) \quad \text{(almost everywhere)} \tag{A.4}$$

Therefore,

$$\Pr(X \in A) = E\big(\Pr(X \in A | G(A))\big) = E(G(A)) = \frac{\alpha(A)}{\alpha(\chi)}$$

A.5

According to Bayes' law,

$$\Pr(\theta_i | \theta_{-i}, t_i) \approx \frac{\alpha}{\alpha+n} f(t_i | \theta_i) \Pr(\theta_i | \theta_{-i}) + \frac{1}{\alpha+n} \sum_{j=1, j \neq i}^{n} f(t_i; \theta_j) \delta(\theta_i = j) \tag{A.5}$$

Therefore,

$$P(\theta_i | \theta_{-i}, t_i) = k \left[\frac{\alpha}{\alpha+n} f(t_i | \theta_i) \Pr(\theta_i | \theta_{-i}) + \frac{1}{\alpha+n} \sum_{j=1, j \neq i}^{n} f(t_i; \theta_j) \delta(\theta_i = j) \right] \tag{A.6}$$

If we set $b = \dfrac{k}{\alpha+n}$,

$$P(\theta_i | \theta_{-i}, t_i) = b \left[\alpha f(t_i | \theta_i) \Pr(\theta_i | \theta_{-i}) + \sum_{j=1, j \neq i}^{n} f(t_i; \theta_j) \delta(\theta_i = j) \right] \tag{A.7}$$

To obtain the accurate form of this conditional posterior density function, it is necessary to calculate the following integral:

$$\int_\theta \Pr(\theta_i | \theta_{-i}, t_i) = b \int_\theta \left[\alpha f(t_i | \theta_{-i}) \Pr(\theta_i | \theta_{-i}) + \sum_{j=1, j \neq i}^{n} f(t_i; \theta_j) \delta(\theta_i = j) \right] = 1 \tag{A.8}$$

Therefore,

$$\Pr(\theta_i \mid \theta_{-i}, t_i) = bq_0 h(\theta_i \mid t_i) + b \sum_{j=1, j \neq i}^{n} f(t_i; \theta_j) \delta(\theta - \theta_j)$$

in which

$$h(\theta_i \mid t_i) = \frac{G_0(\theta_i) f(t_i; \theta_i)}{\int_\theta G_0 f(t_i \mid \theta) \mathrm{d}\theta}$$

$$b = \left(q_0 + \sum_{\substack{j=1 \\ j \neq i}}^{n} f(t_i; \theta_j) \right)^{-1}$$

$$q_0 = \alpha \int_\theta G_0 f(t_i \mid \theta) \mathrm{d}\theta$$

A.6

This equation gets by applying relations (4.6) and (4.7). According to equation (5.35),

$$\Pr(n^* \mid \mu, n, d) \propto \Pr(\mu) \Pr(n^* \mid \mu)$$

Now, according to equation (5.33),

$$\Pr(n^* \mid \mu, n, d) \propto \Pr(\mu) c_n(n^*) n! \mu^{n^*} \frac{(\mu+n) \beta(\mu+1, n)}{\mu \Gamma(n)}$$

$$\propto \Pr(\mu) \mu^{n^*-1} (\mu+n) \beta(\mu+1, n)$$

$$\propto \Pr(\mu) \mu^{n^*-1} (\mu+n) \int_0^1 x^\mu (1-x)^{n-1}$$

Appendix B
R Programming Codes

This subsection of the Appendix presents useful codes for drawing graphs or simulating data that have been used throughout the book. As mentioned earlier, the R programming language has been used in this study. Note that in the case of applying a specific package of R, the name of the package will be pointed at the first line of codes. It is worth noting that most programs have been written in a function. Hence, to run the program with other attributes (or parameter values), it is enough to change the value of desired parameters in the function.

B.1: For Figure 2.2
```
t<-seq(0,3,by=.1); nt=length(t)
ft<-matrix(nrow=1); Ft<-matrix(nrow=1)
ht<-matrix(nrow=1); Rt<-matrix(nrow=1)
fu<-function(a,b) {
for (j in 1 : nt)
{
ft[j]<-dweibull(t[j], shape=a,scale=b)
Ft[j]<-pweibull(t[j],shape=a,scale=b)
Rt[j]<-1-Ft[j]
ht[j]<-ft[j]/Rt[j]
}
list(par(mfrow=c(2,2)), plot(ft,type="l",
main="Probability Density Function" ,
xlab="t",ylab="ft"),plot(Ft,type="l",
main="Cumulative Distribution Function" ,
xlab="t",ylab="Ft"),plot(Rt,type="l",
main="Reliability Function" ,xlab="t",ylab="Rt")
, plot(ht,type="l",
main="Hazard Function",xlab="t",ylab="ht"))
}
fu(2,.9)
```
B.2: For Figure 2.3
```
ht<-function(a1,a2,a3){
y1<-matrix(nrow=1); y2<-matrix(nrow=1)
y3<-matrix(nrow=1); t<-seq(0,10,by=.1); b<-.9
for(i in 1: length(t)){
y1[i]<-a1*b*t[i]^(a1-1)
y2[i]<-a2*b*t[i]^(a2-1)
y3[i]<-a3*b*t[i]^(a3-1)  }
list(par(yaxt="non"),plot(t,y1,type="l",lty=2,lwd=3,
ylab="ht",xlab="t",main="Hazard Function")
,par(new=T),plot(t,y2,type="l",lty=9,lwd=3,
```

```
ylab="ht",xlab="t",main="Hazard Function")
,par(new=T), plot(t,y3,type="l",lty=1,lwd=3
,ylab="ht",xlab="t",main="Hazard Function")) } ht(.5,1,3)
legend(3,190,legend=expression("shape parameter"==.5,
"shape paremeter"==1,"shape parameter"==3),lty=c(2,9,1),bty
="n")
```

B.3: For Slice Sampling

```
NoSC<- 0 Number of calls of the slice sampling function
NoDE<- 0 Number of evaluated densities during these cells
U.Slice <- function(x0, g, w=1, m=Inf, lower=-Inf,
upper=+Inf, gx0=NULL)
{
#Check the validity of the arguments.
if (!is.numeric(x0) || length(x0)!=1
|| !is.function(g)
|| !is.numeric(w) || length(w)!=1 || w<=0
|| !is.numeric(m) || !is.infinite(m) (m<=0 || m>1e9 ||
floor(m)!=m)
|| !is.numeric(lower) || length(lower)!=1 || x0<lower
|| !is.numeric(upper) || length(upper)!=1 || x0>upper
|| upper<=lower
|| !is.null(gx0) (!is.numeric(gx0) || length(gx0)!=1))
{
stop ("The Slice Sampling Argument is Invalid")
}
```

Count the number of calls of this function.

```
NoSC<<- NoSC+ 1
```

the amount of the log density at the initial point

```
if (is.null(gx0))
{ NoDE<<- NoDE+ 1
gx0 <- g(x0)
}
```

Determining the level of slice

```
logy <- gx0 - rexp(1)
```

Finding the initial interval for sampling

```
u <- runif(1,0,w)
M1 <- x0 - u
M2 <- x0 + (w-u) should guarantee that x0 is in [M1,M2]
```

Expand the estimated interval until its ends are outside the slice, or until the limit on steps is reached.

Appendix B

```
if (is.infinite(m)) no limit on number of steps
{
repeat
{ if (M1<=lower) break
NoDE<<- NoDE+ 1
if (g(M1)<=logy) break
M1 <- M1 - w
}
repeat
{ if (M2>=upper) break
NoDE<<- NoDE+ 1
if (g(M2)<=logy) break
M2<- M2 + w
}
}
else if (m>1) limit on steps, bigger than one
{
M3 <- floor(runif(1,0,m))
M4 <- (m-1) - M3
while (M3>0)
{ if (M1<=lower) break
NoDE<<- NoDE+ 1
if (g(M1)<=logy) break
M1 <- M1 - w
M3 <- M3 - 1
}
while (M4>0)
{ if (M2>=upper) break
NoDE<<- NoDE+ 1
if (g(M2)<=logy) break
M2 <- M2 + w
M4 <- M4 - 1
}
}
```

Shrink interval to lower and upper bounds.

```
if (M1<lower)
{ M1 <- lower
}
if (M2>upper)
{ M2 <- upper
}
```

Sample from the interval, shrinking it on each rejection.
repeat

```
{
x1 <- runif(1,M1,M2)
NoDE<<- NoDE+ 1
```

```
gx1 <- g(x1)
if (gx1>=logy) break
if (x1>x0)
{ M2 <- x1
}
else
{ M1 <- x1
}
}
```

Return the point sampled, with its log density attached as an attribute.

```
attr(x1,"log.density") <- gx1
return (x1)
}
```
B.4: For Figure 3.1
```
library(DirichletReg)
Dirichlet<-function(n){
a<-c(.1,1,10)
g1<-rdirichlet(n,c(a[1],a[1],a[1]));
g2<-rdirichlet(n,c(a[2],a[2],a[2]))
g3<-rdirichlet(n,c(a[3],a[3],a[3])); g4<-rdirichlet(n,a)
list(par(mfrow=c(2,2),xaxt="non",yaxt="non"),
plot(g1,col=6,main="alpha=(0.1,0.1,0.1)")
,plot(g2,col=6,main="alpha=(1,1,1)"),
plot(g3,col=6,main="alpha=(10,10,10)")
,plot(g4,col=6,main="alpha=(0.1,1,10)"))
}
Dirichlet(1000)
```
B.5: For Figure 3.2
```
library(DirichletReg)
Dirichlet<-function(n){
a<-c(.1,1,10)
g1<-rdirichlet(n,c(a[1],a[1],a[1]))
g2<-rdirichlet(n,c(a[2],a[2],a[2]))
g3<-rdirichlet(n,c(a[3],a[3],a[3]))
g4<-rdirichlet(n,a)
y<-1:ncol(g1)
x<-1:nrow(g1)
list(par(mfrow=c(2,2),xaxt="non",yaxt="non")
,image(x,y,g1,main="alpha=(0.1,0.1,0.1)")
,image(x,y,g2,main="alpha=(1,1,1)")
,image(x,y,g3,main="alpha=(10,10,10)")
,image(x,y,g4,main="alpha=(0.1,1,10)"))
}
Dirichlet(1000)
```
B.6: For Figure 3.3
```
library(DirichletReg)
f<-function(n){
y1<-rdirichlet(n,c(.1,.1))
y2<-rdirichlet(n,c(1,1))
```

Appendix B

```
y3<-rdirichlet(n,c(10,10))
list(par(xaxt="non",yaxt="non"),plot(density(y1),
xlim=c(0.04,.96),xlab="x",ylab="Dir(x,a,a)",col=2,
lwd=2,lty=1,main=" Density of Dirichlet Distribution"),
par(new=T),plot(density(y2),
xlim=c(0.1,.9),xlab="x", ylab="Dir(x,a,a)",col=3,lwd=4,lty=3,
main=" Density of Dirichlet Distribution"),
par(new=T),plot(density(y3),xlim=c(0.3,.7),
xlab="x",ylab="Dir(x,a,a)",col=4,lwd=6,lty=6,
main=" Density of Dirichlet Distribution"))
}
f(100000)
legend(.44,2.5,fill=c(2,3,4),legend=expression
("a"==0.1,"a"==1,"a"==10),lwd=c(2,4,6),lty=c(1,3,6),bty="n")
```
B.7: For Figure 3.4
```
library(DirichletReg)
f<-function(n){
a<-c(.1,1,20,80)
x<-seq(0,10,length=n)
g0<-pweibull(x,2,3)
g<-matrix(nrow=n,ncol=4)
for(j in 1:4){
for(i in 1:length(x)){
g[i,j]<-rbeta(1,a[j]*g0[i],a[j]*(1-g0[i]))
}
}
list(par(mfrow=c(2,2),xaxt="non"),
plot(x,g0,xlab="alpha=0.1",ylab="CDF"),par(new=T),
plot(g[,1],col=4,,xlab="alpha=0.1",ylab="CDF",pch=3),
plot(x,g0,xlab="alpha=1",ylab="CDF"),par(new=T),
plot(g[,2],col=4,xlab="alpha=1",ylab="CDF",pch=3),
plot(x,g0,xlab="alpha=80",ylab="CDF"), par(new=T),
plot(g[,4],col=4,xlab="alpha=80",ylab="CDF",pch=3),
plot(x,g0,xlab="alpha=20",ylab="CDF"), par(new=T),
plot(g[,3],col=4,xlab="alpha=20",ylab="CDF",pch=3))
}
f(400)
```
B.8: For Figure 3.5
```
library(Renext)
f<-function(n){
a<-c(.1,1,4,9,15,22,30,50,80)
x<-seq(0,10,length=n)
g0<-plomax(x,3.2,2.5)
g<-matrix(nrow=n,ncol=9)
for(j in 1:9){
for(i in 1:length(x)){
g[i,j]<-rbeta(1,a[j]*g0[i],a[j]*(1-g0[i]))
}
}
list(par(mfrow=c(3,3),xaxt="non"),
plot(x,g0,xlab="alpha=0.1",ylab="CDF"),par(new=T),
```

```
plot(g[,1],col=4,,xlab="alpha=0.1",ylab="CDF",pch=3),
plot(x,g0,xlab="alpha=1",ylab="CDF"),par(new=T),
plot(g[,2],col=4,xlab="alpha=1",ylab="CDF",pch=3),
plot(x,g0,xlab="alpha=4",ylab="CDF"), par(new=T),
plot(g[,3],col=4,xlab="alpha=4",ylab="CDF",pch=3),
plot(x,g0,xlab="alpha=9",ylab="CDF"), par(new=T),
plot(g[,4],col=4,xlab="alpha=9",ylab="CDF",pch=3),
plot(x,g0,xlab="alpha=15",ylab="CDF"), par(new=T),
plot(g[,5],col=4,xlab="alpha=15",ylab="CDF",pch=3),
plot(x,g0,xlab="alpha=22",ylab="CDF"), par(new=T),
plot(g[,6],col=4,xlab="alpha=22",ylab="CDF",pch=3),
plot(x,g0,xlab="alpha=30",ylab="CDF"), par(new=T),
plot(g[,7],col=4,xlab="alpha=30",ylab="CDF",pch=3),
plot(x,g0,xlab="alpha=50",ylab="CDF"), par(new=T),
plot(g[,8],col=4,xlab="alpha=50",ylab="CDF",pch=3),
plot(x,g0,xlab="alpha=80",ylab="CDF"), par(new=T),
plot(g[,9],col=4,xlab="alpha=80",ylab="CDF",pch=3))
}
f(400)
```

B.9: For Figure 3.6

```
cluster<-function(n,alpha){
a1<-.1
x<-matrix(ncol=4,nrow=n+1)
x[1,1]<-x[1,2]<-x[1,3]<-x[1,4]<-a1
cluster.number<-c()
for(j in 1:4){
for (i in 1:n)
if(runif(1,0,1)< alpha[j]/(alpha[j]+i)){
Add a new ball color.
new.beta<-rbeta(1,2,3)
x[i+1,j]<-new.beta
cluster.number[j]<-cluster.number[j]+1
} else {
Pick out a ball from the urn, and add back a ball of the same
color.
select.x<-x[sample(1:length(x[,j]),1),j]
x[i+1,j]<-select.x
}
}
}
list(par(mfrow=c(2,2)),plot(x[,1],xlab="alpha=.1",
ylab="Number of Clusters (Frequency)",col="blue",
pch=2), plot(x[,2],xlab="alpha=1",
ylab="Number of Clusters (Frequency)",col="blue",
pch=2), plot(x[,3],xlab="alpha=10",
ylab="Number of Clusters (Frequency)",col="blue",
pch=2), plot(x[,4],xlab="alpha=100",
ylab="Number of Clusters (Frequency)",col="blue",
pch=2),cluster.number)
}
cluster(30,alpha=c(.1,1,10,100))
```

Appendix B

B.10: For Figure 3.8
```
crp = function(num.customers, alpha) {
table<-matrix(ncol=4,nrow=num.customers)
table[1,1]<-table[1,2]<-table[1,3]<-table[1,4]<-1
next.table<-2
for(j in 1:4){
for (i in 1:num.customers-1) {
if (runif(1,0,1)< alpha[j] / (alpha[j] + i)) {
Add a new ball color.
table[i+1,j]<-next.table
next.table<-next.table+1
} else {
Pick out a ball from the urn, and add back a ball of the same
color.
select.table <- table[sample(1:length(table[,j]), 1),j]
table[i+1,j]<- select.table
}
}
}
list(par(mfrow=c(2,2)),plot(table[,1],xlab="alpha=.1",
ylab="Number Of Clusters",col="blue",pch=7),
plot(table[,2],xlab="alpha=1",
ylab="Number Of Clusters",col="blue",pch=7),
plot(table[,3],xlab="alpha=10",
ylab="Number Of Clusters",col="blue",pch=7),
plot(table[,4],xlab="alpha=100",
ylab="Number Of Clusters",col="blue",pch=7))
}
crp(40,alpha=c(.1,1,10,100))
```
B.11: For Figure 4.1
```
mu<-c (3, 5) ; s2 <-c (0.16, 0.09)
p<-0.4 , n<-100; x <-seq(2,400,by=1)
logx<-log(x) ; nx = length(x)
fx<-matrix(nrow=1); Fx<-matrix(nrow=1)
Rx<-matrix(nrow=1); hx<-matrix(nrow=1)
for (j in 1 : nx)
{
fx [j] <-p * exp(-(logx[j]-mu[1])2/2/s2[1])/
sqrt(2*pi*s2[1])/x[j]+(1-p)*exp(-(logx[j]-
mu [2])2/2/s2[2])/sqrt(2*pi*s2[2])/x[j]
Fx [j] <-p*pnorm(logx[j], mu[1], sqrt(s2[1]))+(1-p)
pnorm(logx[j], mu[2], sqrt(s2[2]))
hx [j] <- fx[j]/(1-Fx[j])
}
Simdata<-cbind(x,Fx,fx,hx)
logt<-matrix(nrow=1)
for (i in 1 : n)
{
u<-runif(1,0,1)
if (u<p)
logt[i]<-rnorm(1,mu[1]; sqrt(s2[1]))
```

```
else
logt[i]<-rnorm(1,mu[2]; sqrt(s2[2]))
}
t<-sort(exp(logt))
list(par(mfrow=c(2,2)),
plot(t,xlab="Percent of t",ylab="t",col="blue")
,plot(fx,ylab="Probability Density Function"
,xlab="t",col="blue",type="l",lwd=3)
,plot(Fx,ylab="Cumulative Distribution Function"
,xlab="t",col="blue",type="l",lwd=3)
,plot(hx,ylab="Hazard Function"
,xlab="t",col="blue",type="l",lwd=3))
```
B.12: For Figures 5.2 and 5.3

#Note that in order to rewrite this programming codes for censored data, the reliabilityfunction must be used in place of the density function.

```
library(MCMCpack)
```

#Set up matrix, initialize

```
st = Number of censored data in each stress level
data1= Data of normal stress level
data<-cbind( All data); t<-data[,1]; Iterations<-10000
n<-length(t)
test<-c(seq(5,50,by=2); seq(60,400,by=20))
test1<-length(test); L<-2000
r<-matrix(nrow=Iterations, ncol=1)
v<-matrix(nrow=Iterations, ncol=1)
fi<-matrix(nrow=Iterations, ncol=1)
nstarSUM<-matrix(nrow=Iterations,ncol=1)
a<-matrix(nrow=1); l<-matrix(nrow=1); beta<-matrix(nrow=1)
astar<-matrix(nrow=1); lstar<-matrix(nrow=1)
betastar<-matrix(nrow=1); nstar<-matrix(nrow=1)
nj<-matrix(nrow=1); S<-matrix(nrow=1, ncol=n); cur.astar<-
matrix(nrow=1)
cur.lstar<-matrix(nrow=1); cur.nstar<-matrix(nrow=1, ncol=1)
cur.nj<-matrix(nrow=1); cur.S<-matrix(nrow=1)
q0j<-matrix(nrow=1,ncol=n-1)
njqj<-matrix(nrow=1); b1<-matrix(nrow=1)
b2<-matrix(nrow=1); b3<-matrix(nrow=1)
u1<-matrix(nrow=1); u2<-matrix(nrow=1)
prob1<-matrix(nrow=1); u0<-matrix(nrow=1)
w0<-matrix(nrow=1); sim.a<-matrix(nrow=1)
sim.l<-matrix(nrow=1); simbeta<-matrix(nrow=1)
C2<-matrix(nrow=1,ncol=n); C3<-matrix(nrow=1)
Z<-matrix(nrow=1); multiZ<-matrix(nrow=1)
W<-matrix(nrow=1)
```

#Set up data

```
ar<-1; br<-0.009976;av<-1; bv<-0.001;afi<-2;bfi<-1.5979
```

Appendix B

```
d<-2; r[1]<-100.24; v[1]<-1;fi[1]<-0.825;
nstarSUM[1]<-n;beta[1]<-rnorm(1,0,100)
for(i in 1:n)
{
a[i]<-runif(1,0,fi)
l[i]<-rinvgamma(1,d,r)
lstar<-l
astar<-a
nj[i]<-1
nstar<-length(astar)
S[i]<-i
}
T<-list()
T1<-list()
T2<-list()
Gibbs Sampling
for(mm in 2:Iterations)
{
for(i in 1:n)
{
k0<-S[i]
k<-nj[k0]
if(k>1)
{
cur.lstar<-lstar
cur.astar<-astar
cur.nstar<-length(cur.astar)
cur.nj<-nj
cur.nj[k0]<-nj[k0]-1
}else if(k==1)
{
cur.lstar<-lstar[-k0]
cur.astar<-astar[-k0]
cur.nstar<-length(cur.astar)
cur.nj<-nj[-k0]
cur.S<-S
k1<-which(cur.S>k0)
cur.S[k1]<-cur.S[k1]-1
}
C<-(d*v[mm-1]*(r[mm-1]∫d))/fi[mm-1]
f<-function(x)(x*(t[i]*exp(st[i]*beta[mm-1]))∫(x-1))
/((r[mm-1]+(t[i]*exp(st[i]*beta[mm-1]))∫x∫(d+1))
In<-integrate(f,lower=0,upper=fi[mm-1])$value
q00<-C*In
for(i0 in 1:cur.nstar)
{
q0j[i0]<-(cur.astar[i0]*exp(st[i]*cur.astar[i0]*beta[mm-1])
(t[i]∫(cur.astar[i0]-1))*exp(-exp(st[i]*cur.astar[i0]*beta[mm1])
(t[i]4cur.astar[i0])/cur.lstar[i0]))/cur.lstar[i0]
njqj[i0]<-cur.nj[i0]*q0j[i0]
}
```

```
b1<-append(q00,njqj[seq(1:cur.nstar)],after=length(1))
b2<-sum(b1[seq(1:cur.nstar+1)])
Find the interval
rmultinom
prob1<-b1/b2
prob2<-rmultinom(1,1,prob=prob1)
prob0<-c(which(prob2==1))
```

#Slice Sampling
#Update all the data for step 1

```
if(prob0!=1)
{
astar<-cur.astar
lstar<-cur.lstar
cur.S[i]<-prob0-1
S<-cur.S
cur.nj[S[i]]<-cur.nj[S[i]]+1
nj<-cur.nj
nstar<-length(astar)
a[i]<-astar[prob0-1]
l[i]<-lstar[prob0-1]
}else if(prob0==1)
{
x01<-runif(1,0,fi[mm-1])
gx0<-NULL
g1<-function(x){ log(x)+(x-1)*log(t[i]*exp(st[i]*beta[mm-1]))-
(d+1)
*log(r[mm-1]+(t[i]*exp(st[i]*beta[mm-1]))∫x) }
new.a<-U.Slice(x0=x01,g=g1,w=1,m=Inf,lower=0,upper=fi
[mm-1],gx0)
new.
l<-rinvgamma(1,d+1,r[mm-1]+(t[i]*exp(st[i]*beta[mm-1]))∫new.a)
cur.astar<-append(cur.astar,new.a,after=length(cur.astar))
cur.lstar<-append(cur.lstar,new.l,after=length(cur.lstar))
astar<-cur.astar
lstar<-cur.lstar
a[i]<-new.a
l[i]<-new.l
k2<-which(astar[]==new.a)
cur.S[i]<-k2
S<-cur.S
cur.nj[k2]<-1
nj<-cur.nj
nstar<-length(astar)
}
}
```

#Step 2: Improve mixing of the chain

```
for(j in 1:nstar)
```

Appendix B

```
{
k3<-which(S[]==j)
u0<-runif(length(k3),0,(t[k3]*exp(st[j]*beta[mm-1]))^astar[j])
w0<-runif(length(k3),0,exp(-(t[k3]*exp(st[j]*beta[mm-
1]))^astar[j]/lstar[j]))
Lower<-max(0,max(log(u0)/log(t[k3]*exp(st[j]*beta[mm-1]))))
Upper<-min(fi[mm-1],min(log(-lstar[j]*
log(w0))/log(t[k3]*exp(st[j]*beta[mm-1]))))
u1<-runif(1,0,1)
invcdf.func<-function(x){ exp(log(x*Upper^(n+1)
+(1-x)*Lower^(n+1))/(n+1)) }
astar[j]<-unlist(lapply(u1,invcdf.func))
lstar[j]<-rinvgamma(1,d+nj[j],r[mm-1]+
sum((t[k3]*exp(d[j]*beta[mm-1]))^astar[j]))
}
```

  #Step 3: update v, phi, beta

```
u2<-rbeta(1,v[mm-1]+1,n)
p<-(av+nstar-1)/(n*(bv-log(u2))+av+nstar-1)
p0<-runif(1,0,1)
if(p0<p)
{
v[mm]<-rgamma(1,shape=av+nstar,scale=1/(bv-log(u2)))
}else if(p0>=p)
{
v[mm]<-rgamma(1,shape=av+nstar-1,scale=1/(bv-log(u2)))
}
r[mm]<-rgamma(1,shape=ar+d*nstar,scale=1/(br+sum(1/lstar)))
u3<-runif(1,0,1)
bfi1<-max(bfi,max(astar))
invcdf.func1<-function(x){ bfi1/exp(log(1-x)/(afi+nstar))}
fi[mm]<-unlist(lapply(u3,invcdf.func1))
u1[mm]<-runif(1,0,exp(astar[mm]*st[mm]*beta[mm-1]))
u2[mm]<-runif(1,0,exp(-(t[mm]^astar[mm])
exp(astar[mm]*st[mm]*beta[mm-1])/lstar[mm]))
l11<-max(log(u1)/(astar*st))
u11<-min(log(-lstar*(t^astar)*log(u2))/(astar*st))
g2<-function(x){log(dnorm(x,0,1000))}
x02=rnorm(1,0,1000)
beta[mm]<-U.Slice(x0=x02,g=g2,w=1,m=Inf,lower=l11,upper=u11
,gx0)
nstarSUM[mm]<-nstar
```

  #Test the result

```
W(the weight parameter)
Z<-rbeta(L,1,v[mm]+n)
multiZ<-1-Z
W<-Z
for(j in 2:L-1)
```

```
{
W[j]<-Z[j]*prod(multiZ[1:j-1])
}
W [L]<-1-sum (W[1:L-1])
for(j1 in 1:n){C2[j1]<-1 }
b3<-append(v[mm],C2[seq(1:n)],after=length(1))
C3<-b3/sum(b3)
for(j in 1:L)
{
prob3<-rmultinom(1,1,prob=C3)
prob4<-c(which(prob3==1))
if(prob4!=1)
{
sim.a[j]<-a[prob4-1]
sim.l[j]<-l[prob4-1]
}else if(prob4==1)
{
sim.a[j]<-runif(1,0,fi[mm])
sim.l[j]<-rinvgamma(1,d,r[mm])
}
}
F(test)
Ftest<-matrix(nrow=n,ncol=L)
Ftest<-matrix(nrow=1,ncol=n)
for(i1 in 1:n)
{
Ftest[i1,]<-1-exp(-t[i1]ˆsim.a/sim.l)
Ftest[i1]<-sum (W*Ftest[i1,1:L])
}
T[[mm-1]]<-Ftest
}
Ft<-matrix(colMeans(do.call(rbind,T)),1,n)
list(par(mfrow=c(2,1)),plot(t,Ft[1,],type="l",col=2,xlab="T
ime
(second)"
,ylab="Failut-time Cdf at 7.1 MV/cm",lty=2,lwd=3)
,plot(ecdf(t),do.points=FALSE,verticals=TRUE,add=TRUE,xlab="T
ime
(second)"
,ylab="Failut-time Cdf at 7.1 MV/cm",lwd=2))
```
B.13

Program codes for estimating the unknown parameters using the parametric method for complete or censored data.

```
library(maxLik) for maximum likelihood
t<-Vector of all data
x<-Vector of number of censored data in each stress level
n<-length(t), j1<-sum(log(t))
logLikFun<-function(param){
alpha<-param[1]
```

Appendix B

```
lambda<-param[2]
beta<-param[3]
(n*x*beta*alpha)-n*log(lambda)+n*log(alpha)+(alpha-1)
j1-(1/lambda)*exp(x*beta*alpha)*sum(f^alpha)
}
mle<-maxLik(logLik=logLikFun,
start=c(alpha=Initial,lambda=Initial,beta=Initial))
summary(mle)
}
```

References

Abbott, L. (1956). Quality and competition: An essay in economic theory. *Science and Society*, 20 (3), 281–283.

Al-Assaf, A.F., & Schmele, J.A. (1993). *The Textbook of Total Quality in Healthcare*, St. Lucie Press, Delray Beach, FL.

Antoniak, C.E. (1974). Mixtures of Dirichlet process with applications to Bayesian nonparametric problems. *The Annals of Statistics*, 6, 1152–1174.

Bai, D.S., & Chun, Y.R. (1993). Nonparametric inferences for ramp stress tests under random censoring. *Reliability Engineering and System Safety*, 3, 217–223.

Bai, D.S., & Lee, N.Y. (1996). Nonparametric estimation for accelerated life tests under intermittent inspection. *Reliability Engineering and System Safety*, 1, 53–58.

Basu, A.P., & Ebrahimi, N. (1982). Nonparametric accelerated life testing. *IEEE Transactions on Reliability*, 5, 432–435.

Beckett, L., & Dlaconis, P. (1994). Spectral analysis for discrete longitudinal data. *Advances in Mathematics*, 1, 107–128.

Bell, S.J., & Halperin, M. (1995). Testing the reliability of cellular online searching. *Online (Wilton, CT)*, 19 (5), 15–24.

Berndt, D.J., Fisher, J.W., & Hevner, A.R. (2001). Healthcare data warehousing and quality assurance. *Computer*, 34 (12), 56–65.

Blackwell, D., & MacQueen, J.B. (1973). Ferguson distributions via Ploya Urn schemes. *The Annals of Statistics*, 2, 353–355.

Cai, K.Y. (2012). *Introduction to Fuzzy Reliability* (Vol. 363). Springer Science & Business Media, Berlin, Germany.

Cassady, C.R., & Kutanoglu, E. (2005). Integrating preventive maintenance planning and production scheduling for a single machine. *IEEE Transactions on Reliability*, 54 (2), 304–309.

Chen, S.M. (1994). Fuzzy system reliability analysis using fuzzy number arithmetic operations. *Fuzzy Sets and Systems*, 64 (1), 31–38.

Chib, S., & Greenberg, E. (1995). Understanding the Metropolis-Hastings algorithm. *American Statistician*, 49, 327–335.

Chittenden, F., Poutziouris, P., & Mukhtar, S.M. (1998). Small firms and the ISO 9000 approach to quality management. *International Small Business Journal*, 17 (1), 73–88. DOI: 10.1177/0266242698171004.

Christensen, R., & Johnson, W. (1988). Modeling accelerated failure time with a Dirichlet process. *Biometrika*, 77, 693–704.

Condra, L. (2001). *Reliability Improvement with Design of Experiment*, CRC Press, Boca Raton, FL.

Congdon, P. (2003). *Applied Bayesian Modeling*, Wiley, New York.

Damien, P., Wakefield, J., & Walker, S. (1999). Gibbs sampling for Bayesian non-conjugate and hierarchical models by using auxiliary variables. *Journal of the Royal Statistical Society*, 331, 344.

El-Aroui, M.A., & Soler, J.L. (1996). A Bayes nonparametric framework for software-reliability analysis. *IEEE Transactions on Reliability*, 45 (4), 652–660.

Escobar, M.D. (1994). Estimating normal means with a Dirichlet process prior. *Journal of the American Statistical Association*, 425, 268–277.

Escobar, M.D., & West, M. (1995). Bayesian density estimation and inference using mixtures. *Journal of the American Statistical Association*, 430, 577–588.

Evans, M., & Swartz, T. (1995). Methods for approximating integrals in statistics with special emphasis on Bayesian integration problems. *Statistical Science*, 10, 254–272.
Faghih, N. (1989). *Maintenance Engineering*, NAVID Publications, Shiraz, Iran.
Faghih, N. (1998). *Statistical Quality Control*, SAMT Publications, Tehran, Iran.
Faghih, N., & Hamedi, M.A. (2007). *Quality and Reliability Engineering*, Farassan Industries Publications, Shiraz, Iran.
Faghih, N., & Loghavi, M. (2007). *Fuzzy Fractal Quality Control*, Rokhshid Publications, Shiraz, Iran.
Faghih, N., & Najafi, Y. (2004). *Fuzzy Reliability in Industrial Systems*, Farabard Industries Publications, Shiraz, Iran.
Faghih, N., & Nobari, N. (2004). *Fuzzy Quality Control*, Farassan Industries Publications, Shiraz, Iran.
Faghih, N., & Yuli, F.A. (2007). *Fuzzy Controllers in Intelligent Material Control Systems*, Rokhshid Publications, Shiraz, Iran.
Faghih, N., & Zadeh, A.E. (2010). *Fuzzy Control in Maintenance Planning*, NAVID Publications, Shiraz, Iran.
Faghih, N., et al. (2013). *Maintenance Planning*, SAMT Publications, Tehran, Iran.
Ferguson, T.S. (1973). A Bayesian analysis of some nonparametric problems. *Annals of Statistics*, 1, 209–230.
Gelman, A., Carlin, J.B., Stern, H.S., & Rubin, D.B. (2003). *Bayesian Data Analysis*, 2nd ed., Chapman Hall/CRC, New York.
Geman, S., & Geman, D. (1984). Stochastic relaxation, Gibbs distribution and Bayesian restoration of images. *IEEE Transactions on Pattern Analysis and Machine Intelligence*, 6, 721–741.
Ghahramani, B. (2003). *An Internet based Total Quality Management System*, Proceedings of the 34th Annual Meeting of the Decision Sciences Institute, pp. 345–349.
Ghosh, S.K., Ghosal, S., Upadhyay, S.K., & Dey, D.K., Eds. (2007). Semi-parametric accelerated failure time models for censored data. In: Upadhyay, S.K., & Dey, D.K. (Eds.), *Bayesian Statistics and Its Applications, Anamaya Publishers*, New Delhi, 213–229.
Giudici, P., Givens, G.H., & Mallick, B.K. (2009). *Bayesian Modeling Using WinBUGS*, John Wiley & Sons, Inc, Hoboken, NJ.
Görür, D., & Rasmussen, C.E. (2010). Dirichlet process Gaussian mixture models: Choice of the base distribution. *Journal of Computer Science and Technology*, 25, 615–626.
Harvey, L., & Green, D. (1993). Defining quality. *Assessment & Evaluation in Higher Education*, 18 (1), 9–34. DOI: 10.1080/0260293930180102.
Hastings, W.K. (1970). Monte Carlo sampling methods using Markov Chains and their applications. *Biometrika*, 57, 97–109.
Juran, J.M. (2004). *Architect of Quality: The Autobiography of Dr. Joseph M. Juran*, 1st ed., McGraw-Hill, New York. ISBN 978-0-07-142610-7, OCLC 52877405.
Kai-Yuan, C., Chuan-Yuan, W., & Ming-Lian, Z. (1991). Fuzzy variables as a basis for a theory of fuzzy reliability in the possibility context. *Fuzzy Sets and Systems*, 42(2), 145–172.
Kalbfleisch, J.D., & Prentice, R.L. (1980). *The Statistical Analysis of Failure Time Data*, Wiley, New York.
Kapur, K.C., & Lamberson, L.R. (1977). *Reliability in Engineering Design*. Wayne State University, Detroit, MI.
Kececioglu, D. (2002). *Reliability Engineering Handbook* (Vol. 1). DEStech Publications, Inc, Lancaster, PA.
Khan, F.I., & Haddara, M.M. (2003). Risk-based maintenance (RBM): A quantitative approach for maintenance/inspection scheduling and planning. *Journal of Loss Prevention in the Process Industries*, 16 (6), 561–573.
Kimberly, J.R., & Minvielle, E. (2000). *The Quality Imperative: Measurement and Management of Quality in Healthcare*, Imperial College Press, London.

References

Kolmogorov, A.N. (1933). *Grundbegriffe der Wahrscheinlichkeitsrechnung.* English transl: *Foundations of the Theory of Probability* (1956), Chelsea Publishing Company, New York.

Korwar, R.M., & Hollander, M. (1973). Contributions to the theory of Dirichlet processes. *Annals of Probability*, 4, 705–511.

Kottas, A. (2006). Nonparametric Bayesian survival analysis using mixtures of Weibull distributions. *Journal of Statistical Planning and Inference*, 136, 578–596.

Kromholtz, G.A., & Condra, L.W. (1993). A new approach to reliability of commercial and military aerospace products: Beyond military quality/reliability standards. *Quality and Reliability Engineering International*, 9 (3), 211–215.

Kuo, L., & Mallick, B. (1997). Bayesian semiparametric inference for the accelerated failure-time model. *Canadian Journal of Statistics*, 4, 457–472.

Lin, C.H., & Kuo, Y. (2011). Single- and dual-layer nanocrystalline indiumtin oxide embedded ZrHfO high-k films for nonvolatile memories—Material and electrical properties. *Journal of the Electrochemical Society*, 8, 756–762.

Lin, Z., & Fei, H. (1991). A nonparametric approach to progressive stress accelerated life testing. *IEEE Transactions on Reliability*, 2, 173–176.

Liu, J.S. (1996). Nonparametric hierarchical Bayes via sequential imputations. *The Annals of Statistics*, 3, 911–930.

Luo, W. (2004). *Reliability Characterization and Prediction of High k Di-electric Thin Film.* PhD Dissertation.

MacEachern, S.N., & Müller, P. (1998). Estimating mixture of Dirichlet process models. *Journal of Computational and Graphical Statistics*, 7, 223–238.

Mann, N.R., Schafer, R.E., & Singpurwalla, N.D. (1974). *Methods for Statistical Analysis of Reliability and Life Data*, Wiley, New York.

Meeker, W.Q., & Escobar, L.A. (1998). *Statistical Methods for Reliability Data*, Wiley, Hoboken, NJ.

Metropolis, N., Rosenbluth, A.W., Rosenbluth, M.N., Teller, A., & Teller, H. (1953). Equations of state calculations by fast computing machines. *Journal of Chemical Physics*, 21, 1087–1091.

Metropolis, N., & Ulam, S. (1949). The Monte Carlo method. *Journal of the American Statistical Association*, 44, 335–341.

Müller, P., Erkanli, A., & West, M. (1996). Bayesian curve fitting using multivariate normal mixtures. *Biometrika*, 1, 67–79.

Müller, P., & Quintana, F.A. (2004). Nonparametric Bayesian data analysis. *Statistical Science*, 1, 95–110.

Neal, R.M. (2003). Slice sampling. *The Annals of Statistics*, 3, 705–767.

Neal, R.M. (2008). *R Functions for Performing Univariate Slice Sampling.* [Online]. Available: https://thatnerd2.github.io/technical/slicesampl.html.

Nelson, W. (1990). *Accelerated Testing: Statistical Models, Test Plans, and Data Analysis*, Wiley, Hoboken, NJ.

Palmer, D. (2006). *Maintenance Planning and Scheduling Handbook* (p. 821). McGraw-Hill, New York.

Reeves, C.A., & Bednar, A.D. (1994). Defining quality: Alternatives and implications. *Academy of Management Review*, 19 (3), 419–445. DOI: 10.5465/amr.1994.9412271805.

Sethuraman, J. (1994). A constructive definition of Dirichlet priors. *Statistica Sinica*, 4, 639–650.

Shaked, M., & Singpurwalla, N.D. (1982). Nonparametric estimation and goodness-of-fit testing of hypotheses for distributions in accelerated life testing. *IEEE Transactions on Reliability*, 1, 69–74.

Singer, D. (1990). A fuzzy set approach to fault tree and reliability analysis. *Fuzzy sets and Systems*, 34 (2), 145–155.

Singpurwalla, N.D. (2006). *Reliability and Risk: A Bayesian Perspective*, Wiley, London.
Smith, A.F.M. (1991). Bayesian computational methods. *Philosophical Transactions of the Royal Society*, 337, 369–386.
Tanner, M.A. (1996). *Tools for Statistical Inference*, 3rd ed., Springer-Verlag, New York.
Tyoskin, O.I., & Krivolapov, S.Y. (1996). Nonparametric model for step-stress accelerated life testing. *IEEE Transactions on Reliability*, 2, 346–350.
Walker, S.G. (2006). Sampling the Dirichlet mixture model with slices. *Statistics - Simulation and Computation*, 36, 45–54.
Walsh, B. (2004). *Markov Chain Monte Carlo and Gibbs Sampling* [Online]. Available: http://nitro.biosci.arizona.edu/courses/EEB596/handouts/Gibbs.pdf
Wu, E.Y., Stathis, J.H., & Han, L.K. (2000). Ultra-thin oxide reliability for ULSI application. *Semiconductor Science and Technology*, 15, 425–435.
Yuan, T., Liu, X., Ramadan, S.Z., & Kuo, Y. (2014). Bayesian analysis for accelerated life tests using a Dirichlet process Weibull mixture model. *IEEE Transactions on Reliability*, 63 (1), 58–67.

Index

accelerated lifetime test 9, 16–17, 57, 59, 77
aperiodic 27–28, 31
average lifespan 1

base distribution 44–48, 51–54, 60, 62, 65–66, 70, 74, 78–79
beta-binomial 34
Blackwell-Macqueen urn scheme 32, 50–53, 56, 68–69, 74
burn-in period 12

Chapman-Kolmogorov equation 25
censored data 9–10, 66–67, 71–72, 82–83, 88, 90, 93, 96, 108, 112
Chinese Restaurant Process 32, 54, 72, 83
conjugate prior 34–35, 43, 46, 79

design engineering 1, 7
Digamma function 56
Dirichlet distribution 32–46, 48–49, 56, 61, 105
Dirichlet process 32–33, 35, 43–54, 56–57, 59–60, 62–64, 66–67, 69–70, 74–75, 77–80, 88–90, 92, 94
Dirichlet process Weibull mixture model 78, 80, 89, 92, 94
durability 5

effectiveness 3
efficiency 3
empirical distribution 48, 77, 90, 95
engineering management 1, 6
estimation of distribution function 90

failure-time distribution 59
flexibility 3, 38, 59, 91, 96
fuzzy control 7

Gibbs sampling 29–30, 32, 67, 69, 73, 80, 85, 96, 109
guarantee 1–2, 16, 102

hazard function 10, 12, 14, 61, 101, 102, 108
hierarchical model 62, 66, 79
hybrid censor 15
hyperparameter 66–67, 69–70, 73, 79–81, 96

importance sampling 9, 22–23, 28
indicator function 36
initial value 28–29, 32, 70
innovation 3
interval censor 15

irreducible chain 27

joint distribution 67, 88

kernel function 73

lifespan 1, 4–5, 8–9, 11, 13, 15, 17
lifetime-stress relationship 18, 90
limiting distribution 25–26, 28–29
location-scale distribution 59
log-linear regression 60, 63, 74, 78
loss function 19, 63, 90

maintenance engineering 6–7
maintenance planning 6–7
marginal distribution 19, 38, 41
Markov chain 9, 20, 23–24, 26–28, 31, 57, 60, 67, 84
Markov Chain Monte Carlo Method 20
maximum likelihood 16, 63, 75, 77, 90, 112
metal-oxide-semiconductor 90
Metropolis-Hastings algorithm 28–30
Monte Carlo Integration 21–22
Monte Carlo method 20–23, 28–29, 32

nonparametric Bayesian approach 43, 59, 66, 91
nonparametric methods 6–7, 63

permutation 30
Polya sequence 50, 52
Pólya's urn model 32, 49–50
Pólya urn process 49
Pólya urn scheme 48, 49
posterior distribution 19–22, 34, 39–40, 43, 48, 51–52, 62, 63, 66–73, 80–83, 85, 88–89, 96
precision parameter 36–38, 42, 44–48, 50–54, 56–57, 60–61, 65–66, 69–70, 73–74, 78–79, 87–88
productivity 1, 3

quality assurance 2, 6
quality control 2, 4, 6, 7, 8, 16
quality improvement 1–2, 8–9, 16
quality management 1–4, 6
quality planning 2
quantile function 14

random censor 15
reliability function 10–11, 13–14, 16, 59, 71, 82, 101
repair and maintenance 1, 7–8
repair capability 5–6

119

scale parameter 13–14, 79
shape parameter 13–15, 78–79, 102
simulation algorithm 47, 55, 57, 60, 63, 74, 80, 96
skew 36
slice sampling 9, 31–32, 82, 96, 102, 110
Stirling number 87
stochastic process 49, 62, 69

survival analysis 7
survival function 10–11

transition probability 23–26

Weibull distribution 13–15, 45–46, 63–65, 79, 90
weight coefficient 74